T0135469

Jan-Gerrit Richter

Fast Measurement of Individual
Head-Related Transfer Functions

Logos Verlag Berlin GmbH

Aachener Beiträge zur Akustik

Editors:
Prof. Dr. rer. nat. Michael Vorländer
Prof. Dr.-Ing. Janina Fels
Institute of Technical Acoustics
RWTH Aachen University
52056 Aachen
www.akustik.rwth-aachen.de

Bibliographic information published by the Deutsche Nationalbibliothek

The Deutsche Nationalbibliothek lists this publication in the Deutsche Nationalbibliografie; detailed bibliographic data are available in the Internet at http://dnb.d-nb.de .

D 82 (Diss. RWTH Aachen University, 2019)

ISBN 978-3-8325-4906-0
ISSN 2512-6008
Vol. 30

Logos Verlag Berlin GmbH
Comeniushof, Gubener Str. 47,
D-10243 Berlin
Tel.:   +49 (0)30 / 42 85 10 90
Fax: +49 (0)30 / 42 85 10 92
http://www.logos-verlag.de

Fast measurement of individual Head-Related Transfer Functions

Von der Fakultät für Elektrotechnik und Informationstechnik der
Rheinischen-Westfälischen Technischen Hochschule Aachen
zur Erlangung des akademischen Grades eines

DOKTORS DER INGENIEURWISSENSCHAFTEN

genehmigte Dissertation

vorgelegt von

Dipl.-Ing.

Jan-Gerrit Richter

aus Bochum, Deutschland

Berichter:

Universitätsprofessorin Dr.-Ing. Janina Fels
Universitätsprofessor Dr.-Ing. Peter Jax

Tag der mündlichen Prüfung: 22.02.2019

Diese Dissertation ist auf den Internetseiten der Hochschulbibliothek online verfügbar.

Abstract

While binaural technology applications gained in popularity in recent years, the majority of applications still use non-individual Head-Related Transfer Functions (HRTFs) from artificial heads. These datasets enable a reasonably good spatial localization which works especially well when using an additional visual cue.

However, certain applications, for example research of spatial hearing or hearing attention, require an physically exact and realistic binaural signal. Moreover, it was shown in many experiments that there is a substantial gain from the use of individual HRTFs, for example in localization tasks. The limiting factor that prohibits the widespread use of individual HRTFs is the acquisition of such data. A substantial hardware requirement obstructs a more universal usage. Even for institutions that allow individual measurements, the measurement time that is required, and that the subjects are required to remain motionless made most measurements unfeasible in the past. This time requirement has recently been reduced by the use of parallelization in the measurement signal which lead to the development of fast measurement systems capable of acquiring individual and spatially dense HRTF.

This thesis provides a objective and subjective evaluation of such a system that is designed with the goal of little disturbance of the measurements in mind. The construction is detailed, followed by both an objective and subjective evaluation. A detailed investigation into additional distortion of the sound field introduced by the system itself is presented and it is shown that the system performs comparably to a conventional system in terms of sound source localization.

Furthermore, a method is introduced and evaluated to further reduce the measurement time by using continuous rotation during the measurement. This method is used to reduced the measurement duration from eight minutes to three minutes without audible differences. The introduced methods are also used to reducing additional errors from subject movement. It is shown that this movement can be reduced by a visual feedback system to a level that can be compensated efficiently.

Vorwort

Die vorliegende Dissertation ist das Ergebnis meiner sechsjährigen Arbeit am Institut für Technische Akustik zwischen 2013 und 2018. Diese Zeit wird mir als eine glückliche Zeit in Erinnerung bleiben, welche mich sowohl fachlich als auch menschlich geprägt hat. An dieser Stelle möchte ich mich bei allen Leuten bedanken, die mich in den letzten Jahren unterstützt haben und ohne die diese Arbeit nicht entstanden wäre.

Als Erstes geht mein Dank an Prof. Janina Fels, für die Chance am Institut zu arbeiten, für die Betreuung meiner Arbeit und die Freiheit, die sie mir geboten hat, meine Ideen auszuprobieren. Des Weiteren danke ich Prof. Peter Jax für die Übernahme der Zweitkorrektur meiner Arbeit und den stets schnellen und unkomplizierten Umgang. Prof. Vorländer möchte ich für den Aufbau der von mir genutzten Strukturen danken. Sowohl die Hardware als auch Software des Instituts haben meine Arbeit erst ermöglicht.

Ein besonderer Dank geht an Gottfried Behler, der mir vor neun Jahren meinen ersten Hiwi-Job am ITA gegeben hat und der mir seitdem immer für Fragen offen stand. Viele fruchtbare Diskussionen und Projekte haben mein Akustik-Wissen immer wieder auf die Probe gestellt und mich, über den Bereich meiner Dissertation hinaus, bereichert.

Einen wichtigen Grundstein meiner Arbeit bildet der Messbogen. Dieses System und zahlreiche andere wären ohne die elektrische und die mechanische Werkstatt des Instituts nicht möglich gewesen. Für die unermüdliche Arbeit, von der alle Mitarbeiter des Instituts profitieren, möchte ich Rolf Kaldenbach, Uwe Schlömer und deren Teams danken. Des Weiteren danke ich Karin Charlier für die Unterstützung bei allen bürokratischen Angelegenheiten.

Allen Assistenten und Studenten des Instituts gilt mein besonderer Dank. Durch Euch gibt es am Institut nicht nur sehr gute Zusammenarbeit, sondern auch manchmal dringend benötigte Motivation und Freundschaften. Besonders möchte ich Josefa Oberem danken, die mich zuerst überzeugt hat, am Institut zu bleiben, und sechs Jahre lang eine gute Büro-Nachbarin und Freundin war. Außerdem danke ich für viel Unterstützung in meinen ersten Jahren Dr.-Ing. Martin Pollow, Dr.-Ing. Markus Müller-Trapet und Dr.-Ing. Martin Guski. Für immer unterhaltsame, dringend benötigte Kaffeepausen und gute Zusammenarbeit danke ich außerdem Michael Kohnen, Jens Mecking, Phillip Schäfer und Hark Braren.

Etliche Versuche, welche in dieser Arbeit vorgestellt werden, wurden von Studenten durchgeführt und ausgewertet. Für ihren Einsatz danke ich Angela Friedrich, Dorothea Setzer, Shaima'a Doma und Saskia Wepner.

Letztlich geht mein Dank an meine Freunde und Familie. Meine Eltern, Großeltern und Geschwister haben mich zu dem gemacht, der ich heute bin, und ich bin sehr dankbar für die anhaltende Unterstützung bei allem, was ich mache. Meinem guten Freund Christian Rohlfing danke ich für die lange Freundschaft, die uns schon seit dem ersten Semester des Studiums verbindet.

Nicht zu vergessen und unschätzbar ist die Unterstützung, welche ich von meiner Freundin Cordula in den letzten Jahren erfahren habe. Auch ihr gilt mein herzlicher Dank. Vieles von dem, was ich in den letzten Jahren erlebt und erreicht habe, wäre ohne sie nicht möglich gewesen.

Contents

1

Introduction

Binaural technology describes the processes required to synthesize spatial sound sources to the listeners eardrums. The spatial location of any sound in a space around the listener is perceived based on differences between both ears both in the time and the frequency domain. These differences can efficiently be described using filters which are called the Head-Related Transfer Functions (HRTFs). For simplicity reasons so called *non-individual* HRTF are commonly used. These functions make use of an artificial head which is an anatomical approximation of a typical human head and usually has microphones integrated in the ear canals. The HRTF of artificial heads aim to perform reasonably well for all subject and are more easily measured even for long durations. Non-individual HRTF have been proven useful for a variety of applications which explains their widespread use. In recent years however, *individual* binaural technology gained more and more popularity. In contrast to non-individual technology, these methods are based on the actual, individual differences for each subject.

The increased focus on individual technology is motivated by shortcoming in the binaural quality from non-individual HRTFs. Listening experiments performed with real sound sources and non-individual HRTF revealed a significant loss in localization performance almost 50 years ago (e.g. [1]). In modern research, the use of individual HRTFs are an important prerequisite for many applications. This is, for example, the case for all experiments involving children. It was shown that the HRTF for children significantly differ from adults HRTFs [2]. However no artificial head with corresponding geometries is available, making individual measurements necessary. Furthermore, there are many experiments where it was shown that the use of individual data results in improved results. The area where this is most researched is the localization of sound sources where individual HRTFs perform better than non-individual ones [3, 4]. One other, current example is the study of auditory attention [5]. For experiments aiming to study auditory processes, reproducible real world performances can only be achieved with good individual binaural data.

Another contributing factor in the gained popularity of binaural technology is the trend towards virtual reality devices, so called Head Mounted Displays (HMDs), which require the user to wear a display in front of their eyes. This reduces one of the most obvious drawbacks for binaural reproduction – the need to constantly wear a headphone: As users already wear a device over the eyes, additional headphones are not perceived as a further nuisance and do not break immersion of the experience. The use of *individual* binaural technology in this field is, if nothing more, a selling point to stand out from competitors as the additional visual cues provided by the display, and the use of head movement tracking already improve the quality of the binaural synthesis [6–8].

The necessity to improve binaural synthesis quality motivated the research in two different directions. Firstly, approaches towards shorter measurement time to make individual measurements feasible, and methods for artificially improving non-individual data on the other hand. This thesis deals with the first approach – the reduction of measurement time needed for HRTF acquisition. While the acquisition of individual HRTFs for a limited spatial resolution has been performed for many decades, many mentioned applications usually require a spatial highly dense dataset to react to the subjects head movement or to produce a complex room simulation. The measurement of spatially dense, individual HRTFs however, required a significant reduction in measurement time.

This thesis describes the construction and evaluation of a system designed to acquire spatially highly dense individual HRTF in a short time duration. The time reduction is achieved by the use of the multiple exponential sweep method. While Chapter 2 gives a short overview over the required theoretical background, Chapter 3 deals with the construction of a measurement system used and evaluated in following chapters. In Chapter 4 an objective evaluation of the influence of the measurement system itself with the use of multiple measurements is presented. Chapter 5 summarizes multiple subjective localization experiments performed to evaluate the validity of the acquired HRTF. An additional improvement which further reduces measurement duration is introduced in Chapter 6. The chapter also includes an objective evaluation of the introduced method. The used methods are further used in Chapter 7. The chapter deals with the influence of subject movement during measurement. An extensive study investigates different measurement methods with regard to the amount of movement the subjects do during a full HRTF measurement.

2

Fundamentals

This chapter gives a brief introduction into the fundamentals used in this thesis. The theory of binaural hearing and the Head-Related Transfer Function (HRTF) are introduced and their applications are explained in detail. Furthermore, basic principles of Linear Time-Invariant (LTI) system identification as well as used interpolation methods are given.

2.1 Binaural Hearing

Humans can intuitively localize sound sources in everyday situations with remarkable precision even without the help of visual stimuli. Binaural hearing, or hearing with two ears, is the underlying requirement for such sound localization. Its basic principles and applications are described in this chapter.

Lord Rayleigh [9] did extensive experiments on the ability of humans to localize sound sources. He proposed that the brain uses differences of both time and level-dependent frequency information between the two ears to extract the location information. These differences are called *binaural cues* in the following. This mechanism can be exploited to trick the brain to perceive virtual sound sources at an arbitrary direction by just adding the corresponding differences to an arbitrary signal, played to the subject. This process, called *binaural synthesis* is the basis for all physically correct virtual acoustic simulation. In the following, the underlying principles are discussed in more detail.

2.1.1 Head-Related Coordinate System

As most of the thesis deals with sound sources relative to the head, a definition of a coordinate system with the head in the center is necessary. The coordinate system is defined as a spherical coordinate system with the head center in the origin. In the following, this center point is defined as the center of the head, halfway between the ears. Figure 2.1 shows the coordinate system definition used throughout this thesis. From the mathematical definition of the spherical coordinate system, two spherical angles, φ and θ and a distance r are defined. While φ corresponds to the azimuth angle, starting at $0°$ in front of the subject and increasing counter-clockwise (mathematically positive), the zenith angle θ is used only in a variation describing the *elevation* and is denoted as φ.

The elevation angle is defined starting at $0°$ in front of the subject, with $90°$ at the top and $-90°$ at the bottom. The transformation from θ to ϑ is described as:

$$\vartheta = 90° - \theta. \tag{2.1}$$

Together with the two angle, three planes are defined [10]. The *horizontal plane*, $\vartheta = 0°$, divides the area around the head into two hemispheres, upper and lower elevation. It also contains the *interaural axis*, $\vartheta = 0°, \varphi = \pm 90°$, which contains both ears. The second important plane is the *median plane*, $\varphi = 0°, 180°$, which divides the area around the head into left and right hemisphere. The *frontal plane* , $\varphi = \pm 90°$ is the third plane, dividing the area around the head into front and back hemispheres.

2.1.2 Binaural Cues

As discussed by Lord Rayleigh [9], two kinds of binaural localization relevant cues can be distinguished. The first cue is described as the difference in time of arrival of one signal between the two ears. The time differences, also called Interaural Time Differences (ITDs), are caused by the a finite propagation speed of sound and the distance of the two ears. Depending on the location of the source, the time difference differs, being almost zero for direction in the front or back of the subject, and having maxima at both sides.

Rayleigh's duplex theory states that these difference are dominant to localization only at low frequencies. At frequencies starting at $f \approx 1500 Hz$, the wavelength of the incident wave are comparable to the size of the head and the differentiation

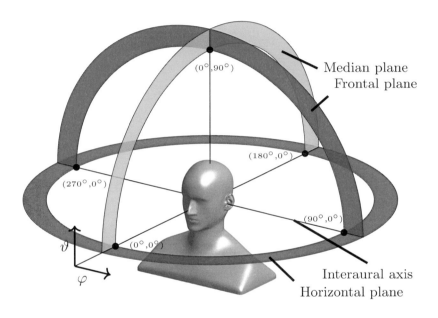

Figure 2.1: Head-related coordinate system definitions.

between the phase of the ears signals is ambiguous. This phenomenon is similar to typical sampling problems where at a certain frequency aliasing effects arise if the sampling is too sparse. While these findings are valid for pure-tones, more recent studies showed that subjects are able to extract time information from the envelopes of higher-frequency stimuli if multiple frequencies are contained [11–13]. At higher frequencies, as the wave is more and more affected by the listeners head, a different binaural cue can be used to localize the sound source. This cue is based on this effect of the head on the sound wave. The higher the frequency, the more the wave is affected by shadowing through reflection, absorption of sound energy and diffraction on the surfaces. Together, significant frequency-dependent level differences between the ears are caused. These level differences, called the Interaural Level Difference (ILD), are highly directional-dependent as the shape of the torso, the head and the pinna play an important role for the generation of the cue.

2.1.3 Monaural Cues

As noted by Hornbostel and Wertheimer in 1920 [14], there can be ambiguity in the interaural differences for multiple points. They describe cones where both ITD and ILD share the same values for all points on the cone. One typical example of such a *Cone of Confusion* is the median-plane where ITD and ILDs are at a minimum. To enable localization on these cones, humans rely on monaural cues [15]. These cues are also a result of varying reflection, absorption and diffraction of the head surface, resulting in different spectral coloration for different directions.

2.2 Head-Related Transfer Function

As described above, humans use differences between the two ear signals to localize sound sources. There are three different cues that are used during this localization process. All binaural and monaural cues can be combined as the influence of the head on a propagating wave in the free field. The influence can be regarded as a directional and frequency-dependent filter of a LTI system (for a static scenario without movement of the source or subject). These filters are commonly called HRTFs. Blauert gave the definition for measurements of HRTFs as the transfer functions from the source to the two ears divided by the transfer function from the source to the head-center point of the subject, with the subject missing. This

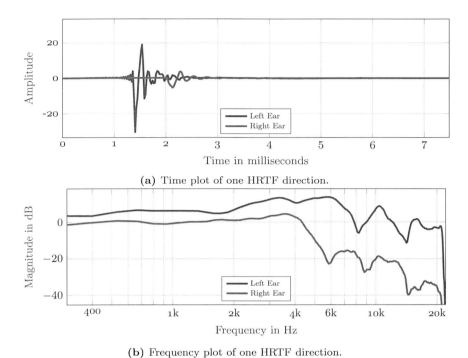

(a) Time plot of one HRTF direction.

(b) Frequency plot of one HRTF direction.

Figure 2.2: An exemplary HRTF for direction $\varphi = 90°, \vartheta = 0°$ from the artificial head in time and frequency domain.

definition of so called *free-field HRTFs* [15] is used throughout this thesis. Figure 2.2 shows an exemplary HRTF for one direction in both time and frequency domain. In the time domain, the ITD can be seen as the impulses are time delayed by approximately 1 ms. The frequency domain plot of the HRTF magnitude shows the frequency-dependent ILD.

2.2.1 Binaural Synthesis

As the brain perceives a source direction merely by processing the differences between the signals at the ears, virtual sound sources can be placed at one desired direction simply by presetting such differences in the presented stimuli. The listeners brain would then perceive the source at the desired direction.

This can be achieved in two ways. The first approach is to use so-called *binaural recordings*. For these recordings, microphones can be inserted into the ears of one person. The resulting recordings contain all spatial cues for localization. This enables the listener to experience a three dimensional recording with sound sources located exactly where they were positioned during the recording. The subject can not, however rotate their head inside the virtual scene or move around as the recording is limited to the microphone positions.

Reports of this method which uses either a two microphone setup, or with the use of artificial heads, described in a later section, can be found in the late 19th century. For a comprehensive historical summary, please refer to Wade and Deutsch [16] or Paul [17].

The second way to create binaural stimuli is called *binaural synthesis* and is achieved using the HRTF filter. To create a binaural signal using binaural synthesis, a mono signal is convolved with the HRTF of one direction of a HRTF dataset [18]. The resulting signal contains all information required to localize the now virtual sound source.

Additionally, two basic reproduction modes are considered in this thesis. A *static* reproduction of a binaural scene is an easy to implement form of binaural synthesis. Here, only the directions of all sources relative to the subject in the scene are needed. Multiple sources can be convolved with the corresponding HRTF of the directions and are summed up to create the full virtual scene. However, the sources are placed in a coordinate system relative to the subjects head. This means that if the subject moves, the sources move with the subject and stay at the same relative position. This presentation mode is as such not a natural reproduction of an everyday acoustic situation and there is a mismatch between expected behavior and perceived signal.

A more natural reproduction method is called the *dynamic* reproduction in this thesis. For this method, additional information about the movement of the subject is required. This information is used to counter all movements in orientation and position by the subject by changing the HRTF filter for each source, depending on the movement. The achieved effect is a source that does not move relative to a global coordinate system. This method requires the capabilities to change HRTF filter in real time [19] and a spatially dense HRTF dataset to be able to place the source at arbitrary positions.

Lastly, the playback of the resulting binaural stimuli needs to be considered. As the stimuli are intended to be perceived directly at the eardrums, without additional filtering, one straightforward method to present these stimuli to the subject is with the use of headphones. For this method, only the transfer function between headphone and eardrums, also called the Headphone Transfer Function (HpTF), needs to be considered as shown in the following.

All uses of binaural synthesis in this thesis is limited to this headphone reproduction as it introduces the least additional uncertainty into the transfer path. Other methods using loudspeakers are available but are not detailed in this work [10].

Headphone Transfer Function

As mentioned before, the binaural signals created by binaural synthesis are intended to reach the subject's eardrums exactly as generated. The easiest way to achieve playback to both ears separately is by the use of headphones. However, each headphone has a unique transfer function that is depending for example on the type of headphone enclosure and size and the used driver. For circumaural headphones, the pinna of the subject also causes disruptions in the path of the soundwave from the headphone driver to the eardrums. Similar to HRTFs, these disruptions are caused by diffraction, refraction and absorption on the pinna and are highly frequency-dependent. Additionally, as each subject's pinna are differently shaped, the transfer function are also subject-dependent.

To achieve a perfect binaural reproduction, the above mentioned individual influences from the headphone, in the following called the Headphone Transfer Function (HpTF), need to be equalized. In this thesis, this equalization is performed using a method introduced by Masiero and Fels [20]. The method uses repeated measurements while letting the subject reposition the headphone each time between measurements. The equalization curve is calculated from the mean in the frequency domain of all measurements to account for changes caused by the fitting of the headphone on the head, as well as individual and system caused differences of the headphone. Figure 2.3 shows an exemplary measurement of ten repetitions of one subject in one ear for a Sennheiser HD650 headphone.

During the individual measurements of the HpTF, but also during individual HRTF measurements, the microphones are placed at the entrance of the subject's ear canal. The fitting of the microphone can either be partly open which occludes the ear canal only from the microphone itself and from a silicon carrier, or closed, with the ear canal fully closed from an ear plug in which the microphone is put on the outside. Both fittings are shown in Figure 2.4. It was shown that both

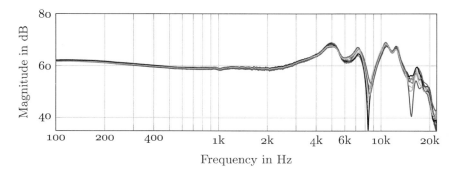

Figure 2.3: Illustration of changes from headphone fitting on the head from 10 repositionings on one subject.

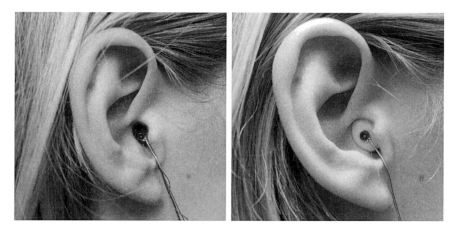

Figure 2.4: Fitting of a miniature microphone with either an open ear canal (left) or closed using an ear plug (right) [21].

fittings are not affecting the authenticity of binaural reproduction compared to real sources if both HpTF and HRTF measurements are performed with the same setting [21].

2.2.2 Artificial Heads

For many binaural applications, it is often easier to use an *artificial head* instead of a human subject with microphones in their ears. These artificial heads, originally conceived as an aid for the development and improvement of hearing aids, mimic a typical human head and torso and have microphones fitted in the artificial ear canals [22]. Several approaches have been made over the years with regards to shape and detail of the head itself.

From these artificial heads, a highly dense HRTF can be acquired as measurements with artificial heads are not limited in time. Additionally, the measurements are highly reproducible and are therefore used extensively to compare different measurement approaches in this thesis. The artificial head used in this thesis was built to resemble one human head, that was selected from a listening experiment with multiple HRTFs [23].

2.2.3 Individual HRTFs

The spatial cues of the HRTF, especially the ILD, are not only highly dependent on the source direction and the frequency, but are a result of the shape of the head and pinna. As these vary substantially between subjects, the HRTF also has a strong individual dependency [24]. The inter-subject differences are mostly located at higher frequencies, where small differences in geometry have a large influence on the frequency and shape of a resonance dip.

Using a HRTF different from your own, by using binaural synthesis with an HRTF dataset from a different head, can greatly affect the quality of the binaural synthesis, as the spatial cues from the used HRTF do not match one person's own individual characteristics. Using a mismatched HRTF, that is a HRTF acquired from a different head, results in a deterioration in localization ability [25]. Additionally, an increase in front-back confusion can be observed [25–27]. This is described in more detail in Section 2.4.

2.2.4 Diffuse-Field HRTFs

The information contained inside the HRTF can be further split into a directionally-dependent and an directionally independent part. The directionally independent part, also called the *diffuse-field HRTF*, can be considered as the result of a measurement of a HRTF not in the free, but in the diffuse field, were no plain wave propagation is possible but instead, waves come at random from all directions at the same time. It is also possible to calculate the diffuse-field HRTFs from an existing set of HRTFs by integrating the HRTFs over all spatial directions [28]:

$$|\mathrm{HRTF}_{\mathrm{diff}}(f)|^2 = \frac{1}{4\pi} \int_\theta \int_\varphi |\mathrm{HRTF}(\theta, \varphi, f)|^2 \cdot \cos(\theta) d\varphi d\theta. \qquad (2.2)$$

The counterpart to the diffuse field HRTF, the directionally-dependent part, is called the Directional Transfer Function (DTF) [29, 30] and can be obtained by subtracting the diffuse field HRTF from the HRTF

$$\mathrm{DTF}(\varphi, \vartheta, f) = \mathrm{HRTF}(\varphi, \vartheta, f) - \mathrm{HRTF}_{\mathrm{diff}}(f). \qquad (2.3)$$

2.2.5 Near-Field HRTFs

While in previous chapters it was shown that the HRTF is highly dependent on frequency and on individual head geometry, one more dependency is often neglected and is the center of this section, the dependency of the HRTF on the measurement distance. Some early observations have been made by Steward [31, 32] who derived mathematical expressions for both ITD and ILD on a rigid sphere with two ears. His conclusions show that the ILD increases as the source position moves closer to the head, while the ITD is approximately independent of the source distance. This theory was again tested with an updated sphere model by Brungart and Rabinowitz [33] and could be validated using measurements by Duda and Martens [34].

Brungart and Rabinowitz also repeated a similar evaluation using measurements on a sphere and an artificial head [35]. As they claim that previous publications show no large distance variations for distances above 1 m, they limited their investigation to changes in a range between 0.2 m and 1 m.

2.3 Measurement

This section gives a short overview of the commonly used measurement techniques and measurement systems. It introduces a reference measurement system that is used as a standard to compare against. Firstly, a brief introduction into LTI-System measurement fundamentals and impulse response measurement using spectral division is given. This is followed by a short description of HRTF measurement techniques and systems for both non-individual and individual HRTFs.

2.3.1 Linear Time-Invariant-Systems

The definition of an LTI system entails two properties. Firstly, the system needs to behave linear to any given input. This means that all measurements are independent of the sound level of the measurement. Secondly, the system needs to behave time invariant. The obtained measurements must be independent of the time the measurement tool place and should be perfectly reproducible. This is, of course, only an approximation as no acoustical system fulfills the linear and time invariant assumptions. In the context of this thesis, all measurements are based on the assumption of an LTI system.

Figure 2.5 shows an LTI system with impulse response $h(t)$. All signals $s(t)$ are convolved with the impulse response $h(t)$ to generate the signal $g(t)$. In the context of HRTF, an arbitrary signal arrives at the subject, where it is convolved with the HRTF of the source direction to generate a signal at the ear canals.

To identify the impulse response, a known input signal $(s(t))$ and output signal $(g(t))$ are prerequisites. A typical method to obtain the impulse response from the measurement, is to transform both $s(t)$ and $g(t)$ to the frequency domain using the Fourier transformation. In the frequency domain, the convolution becomes a simple multiplication (as depicted in Figure 2.5) which is more easily solved for the sought impulse response, than the convolution. From the transfer function $H(f)$ the impulse response can be obtained by the inverse Fourier transformation. The used excitation signals to obtain the impulse response have a long history and wide range. In this thesis, the excitation signals are limited to exponential sweeps introduced by Farina [36] and described in detail by Müller and Massarani [37].

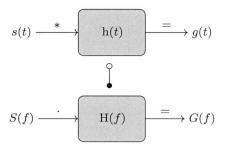

Figure 2.5: Basic principle of the frequency domain analogies of an LTI system.

2.3.2 Measurement of Head-Related Transfer Functions

The typical measurement procedure of HRTFs follows the definition of free-field HRTFs by Blauert [15]. First, the subject is equipped with two miniature microphones, placed inside the ear canal. The subject is then positioned in a full or semi-anechoic chamber with a defined distance and orientation to a loudspeaker. Any typical measurement signal can be used to measure the transfer paths. After the measurement the subject is repositioned and subsequently a HRTF of this new, different direction is measured. This process is repeated until all required directions are measured. Additionally, a reference measurement without the subject is done. During post-processing, all measurements are divided by this reference in the frequency domain.

With this spectral division, the reference has two main functions. Firstly, it removes the initial latency resulting from the measurement distance from all measurements. Secondly, the spectral division removes the influence of both the loudspeaker and microphone transfer function, if slight directional dependencies are disregarded. Differences that result from a different microphone installation situations have to be corrected separately.

At the Institute of Technical Acoustics (ITA), a measurement system, called the *measurement arm* was designed to automate this process. The subject is positioned on a turntable, enabling rotation in azimuth direction while the loudspeaker is fastened to an arm which can also be rotated, enabling changes in elevation. Figure 2.6 shows a typical HRTF measurement setup with the measurement arm and an artificial head. This setup is used in this thesis as a reference setup to compare measurement results against. Depending on the required spatial resolution, this process can be very time consuming. For a full

Figure 2.6: HRTF measurement of an artificial head using the *measurement arm* of the Institute of Technical Acoustics.

measurement with 1° resolution in both azimuth and elevation, over 72 hours are required. It is easily understandable, that no human subjects can be used for these measurements.

The history of individual HRTF measurements is detailed in the following.

2.3.3 Measurement Systems for Individual Measurements

In the beginning of the study of the influence of the individual HRTFs, only single speaker measurements were taken with long measurement durations [38, 39]. The need to increase the measurement speed was apparent as the subjects suffered fatigue during the long measurements which increased unwanted movement between measurements. Over the years, multiple research groups chose typically one of two design options to improve the measurement speed. Indeed, most publications that use individual HRTFs use either a sphere of loudspeakers to acquire the HRTFs for multiple directions without repositioning at all [1, 21], or an arc of loudspeakers using subject [24] or system rotation [27, 40–44]. The measurement signals used in these arrays either did not allow for simultaneous playback of multiple loudspeakers [41], used various pseudo-random noises [40, 42], which do not offer a good performance for time variances, or applied impulse-

Figure 2.7: The first fast HRTF measurement system developed at the ITA by Masiero and Pollow [48].

techniques [1, 41, 44] with a high number of repetitions.

To achieve a shorter measurement duration, Zotkin proposed a reciprocal system [45], which used a spherical array of microphones and employed loudspeakers in the subjects ears, thus removing the need for parallelization of measurement signals altogether. However, this setup suffered from a high frequency cutoff as a result of the size of the loudspeakers, as well as a relatively low Signal to Noise Ratio (SNR), as the measurement signal is played in close proximity of the subject's eardrums.

The need for individual HRTFs with high spatial density lead to the development of systems with an increased number of loudspeakers. Two separate approaches to the measurement system are used in most systems today.

The first approach uses the multiple-exponential sweep method introduced by Majdak [46] which is, together with the optimization done by Dietrich [47], explained in detail below. This measurement signal was used in a measurement system developed at the ITA by Masiero and Pollow [48], shown in Figure 2.7. This system, and a successor to it, is the basis to this thesis and is described in detail in Chapter 3 and evaluated in Chapter 4. An objective evaluation of both system is presented in Chapter 5.

In parallel to the development of the multiple-exponential sweep methods, a second approach to HRTF measurement became popular in recent years. This approach uses adaptive filtering and Least Mean Squares (LMS) techniques for system identification of HRTFs. This approach, proposed by Enzner [49], even enables continuous rotation while generating dense HRTFs with one loudspeaker. This continuous rotation, suggested by Fukudome [50], allows for shorter measurement times, as discussed in Section 6, and was later also suggested for sweep measurements by Pulkki [51] and Dietrich [52] but was in both cases not evaluated. On the basis of the work of Enzner, LMS method acquisition was further improved by Antweiler and Enzner with the use of perfect-sequence measurement signals [53], evaluated and compared to other methods [54] with regards to SNR [55] and studied with regard to multi-channel setups [56]. This lead to the development of a full-spherical HRTF measurement system by Fuss et al. [57].

2.4 Sound Localization

The study of the ability to pinpoint a sound source in space, also called the localization of a sound source, goes back over one hundred years. Early on, the importance of binaural hearing was discovered. The work of Angell and Fite [58] discusses localization experiments on a subject with only one working ear, while the work of Lord Rayleigh [9] extensively studies the capabilities of the estimation of sound direction. Over the years, countless studies on the localization accuracy have been published. The ability to localize sound sources can be split into two parts, the localization accuracy and the localization blur [15]. The localization accuracy describes the ability to point to an absolute position of the sound source, while the localization blur describes the distance two sound sources are allowed to have, to be differentiable in space.

2.4.1 Localization Accuracy

As described above, localization accuracy describes the accuracy of the ability to point to the absolute position of a sound source. There are multiple factors that influence the localization accuracy. Among those factors are the use of individual or non-individual HRTFs, the duration and type of stimulus, static or dynamic reproduction and the pointing method. All factors influence the resulting localization accuracy result. In the following, a short overview of each factor is given.

Localization of real sources

In most experiments, the ability to localize real sources, i.e. coming from physical loudspeakers, is used either to examine the localization ability in general or, in later studies, as a baseline to compare against. Generally, these experiments test the localization ability with the subjects own HRTFs, as the stimulus from the sound source naturally is filtered on its way to the ear canals. If the experiment is performed in an anechoic chamber, the room impulse response can be neglected and pure HRTFs are used. However, depending on the stimulus length and the pointing device, small or big head movements are a concern as subjects automatically move their head to increase localization accuracy. Oldfield and Parker [59] tested eight subjects for localization accuracy. They used a gun pointing method and fixed the head during the stimulus. They found errors of about 5° in the front and 15° in the back for azimuth and a more uniform error of about 8° for elevation. Makous and Middlebrooks tested six subjects with very short noise bursts [60] and head pointing achieving localization errors at of about 2° for the horizontal plane and 3° in the median plane, for sources in front of the subject. For sources in the back, the error was as high as 20°. In the work by Wightman and Kistler [61], test subjects orally reported estimates of the sound sources of either real sources, or headphone synthesized stimuli with individual HRTFs. The reported azimuth angle errors range from 16.1° to 29.8°, depending on the source direction.

Localization using non-individual HRTFs

While most experiments used real sound sources for the general study of the localization ability, with the emerging binaural technology and the development of first artificial heads, the idea to use non-individual HRTFs gained popularity. Many publications point to a degradation of localization performance if non-individual HRTFs are used [3, 4], compared to real sources. Most notably, the localization performance in elevation directions degrades the most. There is also a significant increase in front-back confusion noted by [25–27].

Influence of the Pointing Method

The method, by which the subjects indicate the perceived direction also has a big influence on the results. Over the years, a number of methods have been proposed and evaluated. The methods can be grouped in exocentric methods, in which the subjects are required to indicate the perceived location on a sphere representing auditory space, and egocentric methods, in which the subjects own center is used as a reference.

One of the first and most intuitive ways to report the perceived source direction is by the means of verbal report [62, 63]. While they do not need any additional device present that the subject needs to learn and might cause interference in the test, the concept of reporting azimuth and elevation angles is not intuitive for most subjects [64]. Furthermore, any no-numerical result needs to be converted for any mathematical analysis [65].

Another way of indicating the position is by pointing with some kind of extension of the body. Notable examples are pointing with the head [60, 66, 67], or pointing with the arm/finger or a laser in the hand called manual pointing [68, 69]. Head pointing methods offer the advantage that the coordinate system of the stimulus is concentric with the coordinate system of the pointing method [60], however they do not allow closed-loop experiments where the stimulus plays during the pointing methods, as the subjects need to move their head [70]. Head pointing was also shown to have a slightly higher precision for elevated sources [65]. Naturally, these methods also have their limitations. While some publications note a reduced accuracy for locations behind the subject [60, 62], other researcher report an overestimation of the lateral positions for right-handed subjects for manual pointing [71].

A recent publication by Bahu and colleagues suggests proximal pointing [72] where the subject indicates the position with a pointing device held near his head. It is reported to yield fast response time independent from the angle and easy pointing to any direction if both hands are involved. However, the method also has limits in the back close to the median plane, where the position is difficult to reach. These egocentric methods are generally regarded as more precise than exocentric methods [66, 73, 74]. For this reason, only some examples are given. Notable examples for exocentric methods are the God's Eye View Localization Point (GELP) methods [73] and the Bochum Sphere [75]. Both methods use a sphere on which the subject indicates a position. The methods are generally described as easier to more comfortable and faster than head pointing [70] but at the same time harder to use due to the change in the center of the coordinate system. They are moreover difficult to use in the dark [76]. Alternatively a 3D representation of a sphere on a computer monitor can be used with similar drawback [77–79] in accuracy. For a more detailed summary of pointing methods, please refer to [80].

Localization using dynamic reproduction

The last influence on the localization accuracy is the gain that is expected from the use of small head movements. The experiments by Young [81] were among the first experiments to look at the influence of head movements on front-back confusion. He fixed ear trumpets to rubber tubes connected to the subjects ear canal. As the trumpets were fixed in space, the subjects head movements had no influence on the interaural properties. The subjects reported apparent source positions only in the back for presented click sounds with multiple positions relative to the ear trumpets. The conclusion was drawn that with head movements removed, front-back confusions increased. The connection between head-movements and changes in ITD and ILD was first expressed by Wallach [6] and was reproduced multiple times in later years[7, 8, 78, 82].

2.4.2 Localization Blur

The localization blur is defined as the minimum distance that two sound sources can have to still be differentiated by human subjects as different sources [15]. Typically, the localization blur is measured as a so called Just Noticeable Difference (JND), where two loudspeaker positions are compared to one another. For example, Mills [83] used one loudspeaker on a movable arm around the subject. The loudspeaker produced pure-tones in the range of 250 Hz to 10 kHz for 1 second. As real sources were used, and the stimulus was relatively long, the head was clamped into a fixed position to reduce subjects head movements. The task itself was to report the relative orientation of a second stimulus compared to a first, reference stimulus. Results in the azimuth plane showed a strong dependence on both the stimulus frequency and the azimuth direction of the reference stimulus. While the frequency-dependency does not follow a general trend and merely certain frequency ranges seem to result in a lower JND than others, a very clear exponential increase in JND can be seen in the azimuth. The lowest JND can be seen at the front, with a steady increase towards the sides. A minimum value of around 1 ° is found here while the values reach 10 ° approximately at 85 °. The same experiment regarding JNDs in the median plane only shows results for frequency-dependence in front of the subject. Here, the values range between 1 ° at low frequencies and 4 ° at 2000 Hz.

Blauert lists results of twelve studies on the localization blur [15] in the horizontal plane with different stimuli. Results range from 0.75 ° to a mean of 4 °.

2.5 Interpolation of HRTFs

If the HRTF dataset is not available in the required spatial resolution as, for example, the measurement duration does not allow a dense measurement, the dataset can be interpolated in the spatial domain to generate missing data. However, any interpolation is limited by the number of measured points and is not able to accurately predict the HRTF at any arbitrary direction correctly. The research question that previous research investigated is the audible influence of such interpolation. In the following, two published experiments are summarized exemplary.

Langendijk and Bronkhorst[43] investigated the subjects ability to discriminate virtual sound sources generated from measured and interpolated HRTFs. Six listeners participated in the testing which included a measurement of individual HRTF at 104 positions. From these positions, interpolations to six known loudspeaker positions are calculated from differently distant positions to emulate an interpolation from differently dense measurements. The task itself was a discrimination task in which the subjects had to identify an odd stimuli in pairs of scheme ABAA, BABB,AABA or BBAB.

The results showed that subject where unable to discriminate between measured and interpolated HRTF, if the spatial sampling was $6°$. For a resolution of $20°$subjects could reliably distinguish between measured and interpolated HRTF. They reported both a change in timbre and position of the stimulus.

Minnaar et al.[84] used interpolation in the time-domain from minimum phase HRTFs. A HRTF dataset with $2°$ resolution was used to artificially create lower resolution datasets. Eight subjects were tasked with discriminating sound sources from interpolated and measured HRTF using a Three-Alternative Forced-Choice (3AFC) experiment design. In the study, a strong directional-dependency for the audibility of interpolation between $2°$ and $24°$ was found in listening tests where sound sources at the sides of the subject required the highest spatial resolution to make interpolation inaudible.

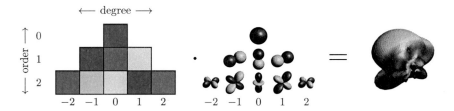

Figure 2.8: Sketch of a spherical harmonic transformation. On the left a depiction of spherical harmonics coefficients multiplied with spherical harmonics basis functions. On the right an example of one frequency of a HRTF of the left ear.

2.6 Spherical Harmonic Decomposition

Analogous to the Fourier transformation, which can describe a 2D wave-signal as a the weighted sum of multiple sine waves with different frequencies [85], the Spherical Harmonic Transformation (SHT) describes spherical objects as a weighted sum of spherical basis functions. Many radiation problems in acoustics can be regarded as a point-source radiation with a spherical directivity pattern. This also applies to HRTFs if the reciprocity principle is applied [45]. Using this principle, the HRTF can be regarded as a frequency-dependent directivity for each ear. The SHT can be used to efficiently describe HRTFs as such [86]. As the SHT is used only as a tool in this thesis, the theory is not given in much detail. A comprehensive elaboration can be found in [87] and [88]. As the sign and normalization can vary, a short overview over the Discrete Spherical Harmonic Transformation (DSHT) and problems related to this thesis is given in the following.

The frequency-dependent directivity g can be described as the weighted sum of spherical harmonic basis functions. The weights a_{nm} are also known as the spherical harmonics coefficients. Equation 2.4 shows the inverse transformation.

$$g(\vartheta, \varphi, f) = \sum_{n=0}^{\infty} \sum_{m=-n}^{n} a_{nm}(f) Y_n^m(\vartheta, \varphi), \qquad (2.4)$$

with a_{nm} as the spherical harmonic coefficients and $Y_n^m(\vartheta, \varphi)$ as the spherical harmonic basis functions with n as the transformation *order* and m are the *degree*. A graphical depiction of the transformation can be seen in Figure 2.8. It should be noted that a dependency on the measurement radius r of the spherical harmonics coefficients is omitted for readability. The used definition of the spherical harmonic basis functions is given analogous to Williams [87] as:

$$Y(\vartheta, \varphi)_n^m \equiv \sqrt{\frac{2n+1}{4\pi} \frac{(n-m)!}{(n+m)!}} \cdot P_n^m(\cos\vartheta) \cdot e^{jm\varphi}, \qquad (2.5)$$

with P_n^m as the associated Legendre function and φ and ϑ as the two room angles.

2.6.1 Order Limit

As seen in Equation (2.4), an infinite order transformation is needed to accurately describe an arbitrary spatial object. As any calculation needs to be finite, an upper order limit needs to be defined. This limit should be chosen to accurately describe full spatial complexity of the spatial object. This order limit N leads to a slightly modified definition seen in Equation (2.6) and a transformation error defined in Equation (2.7).

$$\hat{g}(\vartheta, \varphi) = \sum_{n=0}^{N} \sum_{m=-n}^{n} a_{nm} Y_n^m(\vartheta, \varphi) \qquad (2.6)$$

$$e(\vartheta, \varphi) = \sum_{n=N+1}^{\infty} \sum_{m=-n}^{n} a_{nm} Y_n^m(\vartheta, \varphi) \qquad (2.7)$$

$$g(\vartheta, \varphi) = \hat{g}(\vartheta, \varphi) + e(\vartheta, \varphi) \qquad (2.8)$$

The number of coefficients in a system with maximum order N is calculated as

$$q_{\max} = (N+1)^2 \qquad (2.9)$$

With these definitions, a matrix formulation of the transformation in Equation (2.4) can be given as:

$$\mathbf{g}_a(\vartheta, \varphi, f) = \mathbf{Y} \cdot \mathbf{a}(f), \qquad (2.10)$$

with the base function matrix $\mathbf{Y}$ constructed as a row of base functions for every measurement point.

$$\mathbf{Y} = \begin{pmatrix} Y_0^0(\vartheta_0, \varphi_0) & Y_1^{-1}(\vartheta_0, \varphi_0) & Y_1^1(\vartheta_0, \varphi_0) & \cdots & Y_N^N(\vartheta_0, \varphi_0) \\ \vdots & \vdots & \vdots & \ddots & \vdots \\ Y_0^0(\vartheta_l, \varphi_l) & Y_1^{-1}(\vartheta_l, \varphi_l) & Y_1^1(\vartheta_l, \varphi_l) & \cdots & Y_N^N(\vartheta_l, \varphi_l) \end{pmatrix}. \qquad (2.11)$$

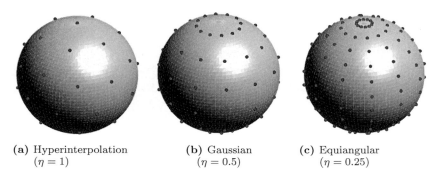

(a) Hyperinterpolation   (b) Gaussian   (c) Equiangular
$(\eta = 1)$ $(\eta = 0.5)$ $(\eta = 0.25)$

Figure 2.9: Examples for three different sampling strategies (order 5) [92].

The matrix $\mathbf{Y}$ is highly dependent on the number of measurements and the measurement positions. As for the transformation, the matrix needs to be inverted, the measurement positions have a big influence on the quality of the transformation itself as discussed in the following.

2.6.2 Sampling Strategies

As described above, the HRTF is described as a spherical function $g(\vartheta, \varphi, f)$ in Equation 2.4. One aspect that was not discussed, is the spatial sampling necessary to obtain this function. With increasing frequency, and decreasing wavelength, the spatial function exhibits a progressively detailed structure. Depending on the distance between two measurement points, these small details can not be described from the measurement points. Their influence is, however, contained in the measurements and is interpreted as energy from lower orders. This phenomenon is called spatial aliasing and is closely related to frequency aliasing effects [89].

To minimize the truncation error and the aliasing effects, specialized spherical sampling strategies have been developed. The hyperinterpolation [90], gaussian [91] and equiangular sampling strategies are just three examples of a specialized sampling. The strategies are developed to ensure an exact transformation up until a maximum order. For each sampling, the efficiency can be calculated as the relation between the maximum coefficients obtained, q_{max} by the use of L measurement points.

$$\eta = \frac{q_{max}}{L}.\tag{2.12}$$

Figure 2.9 shows examples for a hyperinterpolation, gaussian and equiangular sampling scheme for a maximum order of 5. The transformation using any specialized sampling is exact. However, if the sampling is not exact due to, for example, a missing bottom cap, the matrix inversion fails. In these cases, a regularized inversion is needed as described in the following.

2.6.3 Regularization

To calculate the spherical harmonic coefficients from spatial measurement data the inverse of the base function matrix $\mathbf{Y}$ in Equation (2.10) needs to be calculated. The solution of this equation is

$$\mathbf{a}(f) = \mathbf{Y}^{-1} \cdot \mathbf{g}_a(\vartheta, \varphi, f).\tag{2.13}$$

Depending on the measurement positions, this matrix inversion can not be generally calculated. On an ill-conditioned matrix, this would result in a large transformation error. One solution to this problem is to avoid ill-conditioned matrices with special sampling strategies or to employ a *regularization*. This method approximates a solution while balancing an approximation and data error [93]. Duraiswami et al. proposed an order-dependent regularization [94] to invert matrices with high condition numbers. This technique prevents the occurrence of high spherical harmonic orders during the transformation. The SHT in Equation (2.13) becomes

$$\mathbf{a}_{reg} = \left(\mathbf{Y}^H \mathbf{C} \mathbf{Y} + \epsilon \mathbf{D}\right)^{-1} \mathbf{Y}^H \mathbf{C} \mathbf{g},\tag{2.14}$$

with $\mathbf{C}$ as a diagonal weight matrix which should reflect the sphere surface represented by each measurement point, $\mathbf{D}$ as a diagonal matrix with elements $d_j = (1 + i(i + 1))$ and ϵ as the regularization factor.

2.6.4 Acoustic Centering

During the transformation into spherical harmonic coefficients, a higher spatial variance of the input data results into larger orders to be necessary to fully describe the spatial data. As the full complex valued directivity data of the

(a) Directivity plot of a left ear artificial HRTF at 2500 Hz.

(b) Directivity plot of a centered left ear HRTF at 2500 Hz.

Figure 2.10: Influence of the acoustic centering on the phase of a HRTF at 2500 Hz, plotted as a directivity. Magnitude of the HRTF as the radius of the sphere, phase as the color.

HRTF is transformed, the phase of the input data impacts this spatial variance of the data. Figure 2.10a shows an example of a left-ear HRTF at 2500 Hz. The color of the plot represents the phase, while the radius represents the magnitude. As demonstrated by Ben Hagai et al. [95], the energy distribution in the spherical harmonic orders are influenced by the location of the source inside a surrounding spherical array. If the source is centered inside the array, a larger proportion of the energy is located at lower orders. When transforming HRTFs, the receiver inside a surrounding spherical microphone array is never centered as the subject is positioned with its head-center in the array center. Consequently, the ears are offset by half of the heads width towards the respective ear. It was shown that an acoustic centering, accomplished by phase shifting the data by an optimized value towards the ear greatly reduces the maximum transformation order to cover most of the data's energy [86]. As the energy is located at lower orders, the transformation error resulting from the cut of higher orders as well as aliasing effects are reduced during the transformation. A depiction of a centered HRTF directivity plot can be seen in Figure 2.10b. A clear reduction in phase deviation over the sphere is visible. This reduction also reduces the spherical harmonics orders needed. This method is applied to all transformations in this thesis to reduce the influence of the transformation on the data.

3

HRTF Measurement System

This chapter describes the construction of a measurement system to measure spatially dense HRTFs. The goal of this system is to reduce the measurement duration such that individual measurements with a high spatial resolution, which are not feasible with a traditional setup, can be obtained in a reasonable amount of time. As described in Section 2.3.3, several such measurement systems have been designed and constructed in recent years both at the Institute of Technical Acoustics and in institutions all around the world. The goal of the system described in this chapter is to be as small as possible in order to create as little disturbance to the sound field as possible.

The chapter firstly summarizes a previously constructed measurement setup and details drawbacks of the construction that were taken into consideration during construction of the new system. The following part of the chapter details both the loudspeaker housing and the used loudspeakers of the new system before defining the used measurement signal and two different measurement methods that are employed in the rest of this thesis.

3.1 Former Measurement System

This section quickly summarizes the first fast HRTF measurement system build at the ITA previous to the current one. This system is used for some parts of the subjective evaluation in Chapter 5 but ultimately suffered severe hardware damages. These damages lead to the design and construction of the second system whose construction is described in the following. Throughout this thesis, references to this previous system are given.

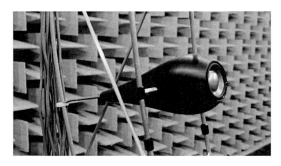

Figure 3.1: Close up view of one loudspeaker in a drop shaped enclosure as employed in the first measurement setup [96].

The system, as seen in Figure 2.7, used 40 loudspeakers on an almost full circle around the subject. The loudspeakers were housed in custom-built drop shaped cases. The cases were selected and tested to have a low influence on the directivity of the speaker and low reflections of neighboring speakers. Figure 3.1 shows one speaker in the drop enclosure. The speakers have been placed on a thin supporting structure. This enabled a free positioning of the loudspeakers on the arc itself with the goal of achieving arbitrary samplings in elevation. The subjects were standing in the center of the construction and were rotated around their center axis. Device rotation could not be achieved using this setup as a consequence of the lightweight design of the truss structure.

During the lifespan of the construction, several drawbacks had been discovered. As the setup was constructed as an almost full circle around the subject (see Figure 2.7), it was rather problematic for the subject to climb into the construction. While this is not a major concern for typical students and adults, the measurement of children and especially elderly would be severely limited.

The main acoustic disturbance caused by the setup was identified as a pivot of the main radiated power towards the side at high frequencies and reflections between neighboring loudspeaker housings, even though the speakers where designed to minimize this effect. While both could be improved by using multiple weighted loudspeakers simultaneously to achieve a good overall directivity [97], this approach resulted in a loss of time efficiency during the measurement.

3.2 Construction

When the previous measurement system suffered severe hardware damages, a construction of a new system was preferred to the repair of the old one as an opportunity to overcome drawbacks from the old construction.

For the new construction, the drawbacks of the old system came into focus. A closed design with one continuous surface is chosen for the new measurement system. The speakers are integrated into the surface. Using this construction, all reflections from neighboring loudspeakers are eliminated and also the placement of more loudspeakers on less space is made possible.

The final construction consists of a vertical, incomplete half-circle construction which houses 64 loudspeakers of 1" diameter. The half circle construction allows for an easy, unhindered access, even with reduced mobility for example for elderly subjects.

The arc itself is constructed as 64 identical segments which are comprised of a small separate and closed volume of 0.05 l per speaker and a rear part which is used as a cable duct as well as two metal pieces that span the whole length of the arc and increase vertical and horizontal stability. Figure 3.2b shows a 3D model of one section without a speaker while 3.2a shows a top view of the segment design. The first volume and the separated second chamber can be seen. Each segment houses one loudspeaker in a fully enclosed volume. The loudspeakers are placed, relative to the center of the arc starting at $\vartheta = -70\,^{\circ}$ and every $2.52\,^{\circ}$ apart till $\vartheta = 88.75\,^{\circ}$ on top of the subject. These unusual distances enables the sampling to function as an equiangular sampling in elevation, as well as a Gaussian sampling of order 90 with only very small positional errors.

The radius of the arc, which also defines the measurement distance, is set to 1.2 m. This was shown to be enough distance to limit audible near-field effects (see Section 2.2.5), while still retaining a good SNR of the acquired HRTF.

Figure 3.3 shows the front view of one segment in the finished arc. The cables are internally routed to the top of the arc, where they can be connected to an amplifier with four Sub-D plugs. This cable management and the connection at the top allows the arc to rotate without twisting internal cables.

Figure 3.4 shows the full new measurement setup from behind. The Figure shows

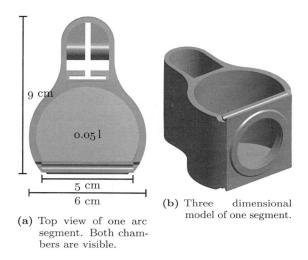

(a) Top view of one arc segment. Both chambers are visible.

(b) Three dimensional model of one segment.

Figure 3.2: Model of one car segment without internals. The frontal chamber houses the speaker and is closed of completely. The second, smaller chamber is used to house wires and supportive structures.

Figure 3.3: Close up view of one of the used loudspeakers in the arc.

a measurement of an artificial head. At the top of the arc, the signal cables can be seen. The closed and rigid construction also enables the full construction to be rotated around the subject instead of using subject rotation. This again lowers the strain on the subject as well as the required physical fitness.

3.2.1 Loudspeakers

The used loudspeakers are of the type *Tang Band W1-2025SA*. They have been selected because of their relatively small size of 1 inch as well as a flat frequency response. Figure 3.5 shows the frequency response in 2 meter distance at 1 Volt. The frequency response can be considered as relatively flat between 500 Hz and 18 kHz. A sharp dip at approximately 15 kHz is also visible. It is assumed that this stems from a resonance in the loudspeaker membrane. This resonance is problematic during the post-processing of the HRTF where a spectral division with the reference is performed. The dip in the reference frequency response increases the energy of the finished HRTF as the dip is slightly frequency- and spatial-dependent and is thus not present in all measurements at the same intensity and frequency. To avoid this behaviour, the reference is spectrally smoothed [98]. The smoothed frequency response is also shown in Figure 3.5.

This resonance is also visible in the directivity plots of the loudspeaker. For this, a spherical cap is measured in front of a single loudspeaker in two meter distance. The cap is measured for $\theta < 60°$ and $0° <= \varphi < 360°$ with the loudspeaker orientated in the positive z axis. This way, an aperture of $120°$ of the directivity was acquired. From this spherical cap, two perpendicular slices are plotted, arbitrarily named *horizontal slice* for $\varphi = 0°, 180°$ and *vertical slice* for $\varphi = -90°, 90°$. As the speaker is round, a perfect piston would not show any differences between the slices.

Figure 3.6 shows two perpendicular slices as isobar plots. These plots show the magnitude of the frequency response, relative to the frontal direction at $\theta = 0°$. Each isobar line represents a change of 3 dB. The plots show the main lobe of the speaker in the center of its directivity. Towards higher frequencies, this main lobe becomes narrower as expected. No side lobes are visible. The vertical slice shows a slightly narrower lobe, indicating that the speakers are not rotationally symmetrical. However, as the narrower lobe remains within 3 dB deviation for $\pm 10°$ until 20 kHz the speakers are suitable for the acquisition of HRTF data. This aperture area is approximately the same size as the surface area of the head

Figure 3.4: Full HRTF measurement setup from behind. Exemplary measurement of an artificial head.

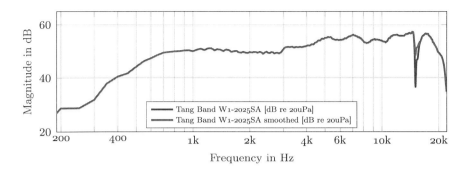

Figure 3.5: Frequency response of the used loudspeaker in 2 Meter distance at 1 Volt before and after smoothing.

and shoulders of the subject. Any deviations outside of this field has only a small influence on the measured HRTF itself.

Notably, the resonance frequency at approximately 15 kHz mentioned in the frequency response is again visible in the isobar plots.

As a comparison to these plots, a directivity of a different loudspeaker of the same size is shown in Figure 3.7. This loudspeaker has been used in a previous version of the measurement setup (see Section 3.1) [99]. While this speaker did not exhibit the resonance behavior, a very pronounced deviation of the energy from the main axis is visible between 12 kHz and 14 kHz. Such a deviating main lobe causes an unequal and uncontrolled sound field at the subject. The rotation of the loudspeaker in the casing is not controlled which causes each deviating main lobe to potentially be directed into a different spatial area. This results in different parts of the subjects geometry to be subjected to more or less energy than intended. This uncontrollable sound field makes the loudspeaker unsuitable for HRTF measurements.

3.3 Measurement Signal

The measurement signal itself has a large impact on any measurement. As described in Section 2.3.3, there are multiple ways to achieve a short measurement time with parallelization of the measurement signals. In this work, all measurements are done with interleaved sweep signals as introduced by Majdak

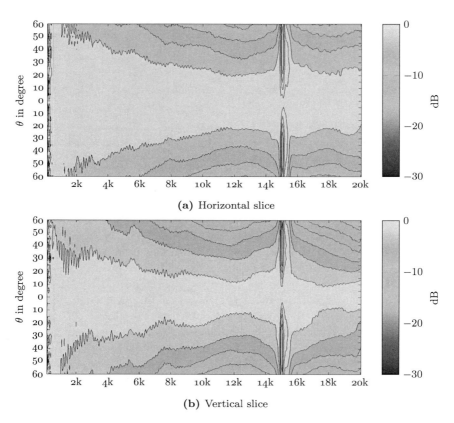

(a) Horizontal slice

(b) Vertical slice

Figure 3.6: Isobar plots of a directivity measurement of the used loudspeaker.

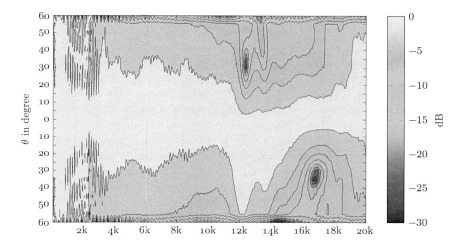

Figure 3.7: Horizontal isobar plot of a directivity measurement of the loud-speakers used in the first measurement setup.

[46] and optimized by Dietrich [47]. The signal consists of multiple exponential sweeps that start with a short time delay t_{wait} relative to each other. This way, multiple sweeps run in parallel at different frequencies. The time for a single measurement of $L = 64$ loudspeakers, each corresponding to one elevation angle, can be calculated as follows:

$$t_{\text{single}} = (\text{L-1})\, t_{\text{wait}} + t_{\text{sweep}} + t_{\text{st}}, \tag{3.1}$$

with t_{st} as the stop margin of the signal, t_{sweep} as the duration of a single loudspeaker sweep and t_{wait} as a time in-between sweep starts of different loud-speakers. The stop margin defines a time in which the system can respond after the last excitation has finished playing. This time is dependent on the full system, specifically on the room in which the measurement takes place. The duration of a single sweep primarily depends on the used frequency rate and the sweep rate which represents the frequency range in octaves divided by the length of the sweep [46]. The optimization by Dietrich [47] returns an optimal sweep rate of $8.6\ \frac{\text{oct.}}{s}$. With this value, and a typically used frequency range between 500 Hz and 22.5 kHz, a single sweep duration can be calculated to $t_{\text{sweep}} = 0.657\,\text{s}$. The delay time t_{wait} is set to 30 ms. A full interleaved sweep of all 64 loudspeakers consequently is of length $t_{\text{single}} = 2.57\,\text{s}$.

3.4 Measurement Types

This section gives detail on two measurement modes that are used in this thesis. The modes concern the subject rotation during the measurement. They are either measured with step-wise or continuous rotation. As the measurement system itself can rotate around one axis, the movement during the measurement can either be from the rotation setup, or from a rotating subject on a turntable with a stationary measurement system. For most considerations, these two options are reciprocal and therefore interchangeable. To improve readability, only subject rotation is discussed in the following; however, all principles also hold for a rotation of the measurement system. All presented methods hold true also for a rotating measurement setup.

3.4.1 Step-wise Measurement

The so-called step-wise measurement describes a more traditional HRTF measurement in which the azimuth positions are measured sequentially. The subject is positioned to one relative azimuth angle and stops there while the measurement runs for all loudspeakers. After the time t_{single}, when the measurement is completed, the subject is rotated so that the next angle can be measured. This process is repeated until all desired measurement positions are covered. The measurement time depends on three parameters. The first is, of course, the measurement time of one sweep of all 64 loudspeakers t_{single} as shown in Equation (3.1). Furthermore, the overall time depends on the time needed for the subject to be repositioned to the next desired angle, called $t_{\text{reposition}}$, and the number of total measurements N:

$$t_{\text{full,stepwise}} = t_{\text{single}} + (N - 1)\left(t_{\text{single}} + t_{\text{reposition}}\right). \tag{3.2}$$

Equation (3.2) shows the time calculation for a step-wise measurement using N azimuth positions. Note that, strictly speaking, the repositioning time $t_{\text{reposition}}$ is also dependent on the number of repetitions N, as for less repetitions the repositioning time increases as a greater angle needs to be covered by the move. Using the step-wise mode, a measurement with a resolution of $5\,^{\circ}$ in φ and $2.5\,^{\circ}$ in ϑ takes approximately 7 minutes. A measurement using the arm setup with the same resolution would take approximately 5 hours. This constitutes a vast reduction in measurement time and makes spatially dense measurements on individual subjects possible.

3.4.2 Continuous Rotation

To further reduce measurement time and subject movement during the HRTF measurement, the continuous rotation is introduced. This measurement mode continuously rotates the subject inside the measurement system, or the system around the subject without stopping.

A substantial decrease in measurement duration can be achieved. This measurement mode is described and evaluated in detail in Chapter 6.

4

Objective Evaluation

While the last chapter described the construction of the measurement system, this chapter contains an objective analysis of the quality of the acquired data. This includes a comparison of measured data to reference measurements and analytical solutions alike.

There are multiple factors that influence a single HRTF measurement. Starting at the measurement chain, the loudspeaker is of course a big influence. A flat frequency response is not as important as a good uniform directivity as already discussed in Section 3.2.1. Amplifiers, digital/analog- and analog/digital-converters introduce latency and their own frequency response, but are assumed to be of high enough quality to be neglected in the following. The microphone transfer functions only play a little role in the finished HRTF measurement, as they are also used in the reference measurement. The microphones are placed at the entrance of the ear-canal with either a blocked or open setting. In the open setting, the ear canal resonance is included in the measurement. However, it was shown that this setting does not affect the plausibility of reproduced sources [21]. Other factors, like the influence of time variances in the measurement environment, that have been shown to have a large impact on room acoustic measurements [100] are avoided by using a reference measurement that is taken a short time before or after the test measurements.

The biggest influences on the measurement are caused by three factors: The disturbance of the sound field by the measurement setup, the position of the subject inside the measurement setup and, in case of individual measurements, subject movements during the measurements. This chapter evaluates the first two factors, while the last one is described in Chapter 7.

The goal of this chapter is to give an overview of the influence of the measurement system itself on the measurement quality, as well as an objective and subjective evaluation of spherical post-processing of the data.

In a first step, objective error measures are defined to quantify a single value error from spherical data. Secondly, in order to analyze the impact of the measurement system, several measurements with objects of increasing complexity are presented.
[1]

4.1 Error Measure

One of the most important tools of any objective comparison is a suitable error measure to compare data. This section gives three error measures used in the following comparison. The main problem when comparing HRTF data is the number of dimensions. There are two ears with both a frequency and directional dependency. A raw comparison, for example a simple difference between measurement and reference, is difficult to visualize. To make the data more manageable, simplification need to be done.

4.1.1 Spectral Differences

In the following, an error measure called *Spectral Differences (SD)* is used. It is based on the inter-subject spectral differences introduced by Middlebrooks [102]. Middlebrooks computes a single value metric by first subtracting the dB values of all frequency bins, component by component, from reference and individual measurements. The variance of the resulting difference over the frequency is calculated and finally the mean of the resulting variances over the whole sphere is computed. For this thesis however, a frequency-dependent metric is desired to, for example, show frequency-dependent influences of the measurement system itself. To this end, the basic idea of the inter-subject spectral difference is used to formulate the Spectral Differences (SD) metric as follows:

$$\text{SD}(f) = \sigma_w \left(20 \log_{10} \left| \frac{\text{HRTF}_1(f, \vartheta, \varphi)}{\text{HRTF}_2(f, \vartheta, \varphi)} \right| \right). \tag{4.1}$$

Equation (4.1) results in a frequency-dependent value that is defined by first taking the ratio of the linear HRTF frequency data (or subtracting them in dB)

---

[1] Parts of this Chapter have been previously published in [101].

and then calculating a standard deviation over the whole sphere for each frequency. Note that for increased readability, the dependency of the ear-side of the HRTF is neglected here. The calculation is done separately per ear. Depending on the used sampling of the HRTF, this standard deviation has to be weighted with spatial weights as the sampling points might not be distributed evenly over the sphere [103]. The spatial weights are calculated by a Voronoi decomposition on the sphere [104]. This way, each error value is weighted with the part of the surface area the measuring point represents. Equation (4.2) shows the definition of the weighted standard deviation with the mean value calculation as shown in Equation (4.3) for N measured points:

$$\sigma_w(x, w) = \sqrt{\sum_{i=1}^{N} w_i \left(x_i - \mu_{\mathrm{w}}\right)^2} \tag{4.2}$$

$$\mu_{\mathrm{w}} = \sum_{i=1}^{N} x_i w_i, \tag{4.3}$$

with w_i as the spatial weights and x_i as the measured values. To obtain differences in phase over the sphere, the logarithmic operation is simply substituted with an argument function

$$\mathrm{SD}_{\mathrm{phase}}(f) = \sigma_w \left(\arg \left(\frac{\mathrm{HRTF}_1(f, \vartheta, \varphi)}{\mathrm{HRTF}_2(f, \vartheta, \varphi)} \right) \right). \tag{4.4}$$

Note that these error measures disregard the mean constant error of the sphere as only variances are evaluated.

4.1.2 Interaural Spectral Differences

The previously introduced measures cover differences in magnitude and phase in a single ear. However, as described in Section 2.2, localization cues can be divided into monaural and binaural cues. To account for differences in binaural cues, the spectral difference measure defined in Equation (4.1) is again slightly modified.

$$\mathrm{IAD}(f) = \sigma_w \left(20 \log_{10} \left| \frac{\mathrm{ID}_1(f, \vartheta, \varphi)}{\mathrm{ID}_2(f, \vartheta, \varphi)} \right| \right), \tag{4.5}$$

with

$$\mathrm{ID}_i = \frac{\mathrm{HRTF}_{i,L}(f, \vartheta, \varphi)}{\mathrm{HRTF}_{i,R}(f, \vartheta, \varphi)}. \tag{4.6}$$

Equation (4.5) defines a modified spectral difference measure for interaural differences. In this measure, not the HRTFs per ear are compared, but interaural differences. To this end, the separate ear data of each data set is first divided by each other to obtain differences between the ears. These differences are then used, analog to single ear data in Equation (4.1), by spectrally dividing and calculating the weighted standard deviation of the logarithmic differences.

Again, a phase relation can simply be calculated analog to (4.4). These four measures cover all monaural and binaural differences in HRTF and are used as an objective measure to quantify measurement insecurities.

4.2 Measurement Setup

This section deals with the evaluation of measurement uncertainties caused by the measurement system itself. The measurement setup is an error source in the sound field, having both reflection and diffraction effects of neighboring speakers. These effects are detrimental to the measurement itself, as HRTF measurement assumes a plane wave.

This section deals with an objective evaluation of the influence of this disturbed soundfield. To this end, several measurements with objects of increasing spatial complexity have been performed which are described in the following.

4.2.1 Directivity

In the first measurement, called the directivity measurement, eight microphones are placed in 5 cm distance to each other, in a horizontal line in the center of the measurement setup. The microphone array is fixed to a stand on a turntable with the center of the array in the rotational center in the middle of the measurement setup. By rotating the array, a circular area with 40 cm diameter was measured. These measurements are used to approximate the sound field in the area that is later occupied by the subjects head. Depending on the used loudspeaker in the arm, the disk is orientated in different angles to the sound wave. The size of the array is chosen to represent a typical size of a human subject inside the array.

Figure 4.1 shows a sketch of the measurement setup with the orientation of the disk in parallel to the sound field for one speaker on ear level as seen from above.

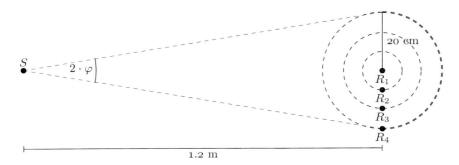

Figure 4.1: Sketch of the measurement setup as seen from above. The impulse response from the source S to a microphone array of four receivers R is measured. The microphones are rotated to cover a full disk. The bold red segment of the outer microphone is used to approximate the directivity in Figure 4.2.

The signal from the outer microphone is used to approximate the directivity of the speaker setup within a limited angle representing the area of the head. Figure 4.2 shows this directivity as an isobar plot with 3 dB distance between levels. The directivity is uniform until approximately 6 kHz. For higher frequencies up until 14 kHz there is some visible distortion at the sides, mostly within 3 dB of the reference direction in $0°$. Between 14 kHz and 18 kHz this distortion is reduced contrary to the expectation. As a reference, Figure 4.3 shows the same section of the directivity of the loudspeaker from the reference measurement system. As expected, this system does not show as many disturbances in the sound field. The plot shows an homogeneous sound field up until 14 kHz, where small errors start to occur at the edges. The direct comparison shows that, while the sound field is more disturbed during the measurement with the fast system, the disturbances lie mostly within 3 dB compared to the reference direction. The measurements are expected to be relatively free of error up until 6 kHz, where some small errors start to occur.

4.2.2 Sphere

As a second test, a solid sphere with diameter 13 cm was used to evaluate the accuracy of the fast measurement system. To this end, the sphere, in which a microphone can be placed such that the microphone sits on the spheres surface, was measured in the reference measurement system described in Section 2.3.2

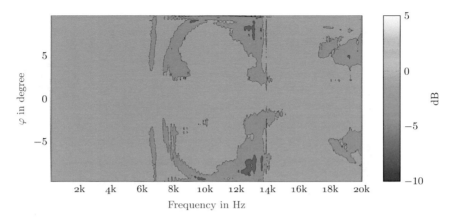

Figure 4.2: Isobar plot of the approximated directivity measurement of one loudspeaker in the measurement system.

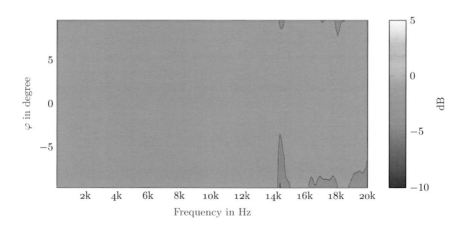

Figure 4.3: Reference isobar plot of the same angle segment shown in Figure 4.2, measured from the loudspeaker used in traditional measurements.

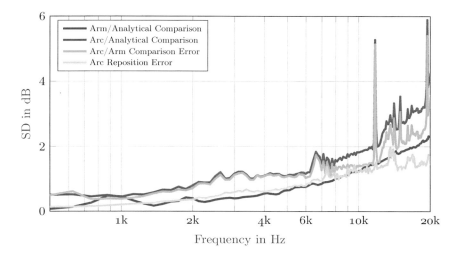

Figure 4.4: Spectral difference measure for two arc and one arm measurements of a solid sphere.

and in the introduced fast system. Additionally, an analytical solution of a solid sphere was created using plane wave scattering on a solid sphere assumptions [105, p. 134]. From this data, both a comparison of inaccuracies of the measurement systems against a simulation and a comparison of the two systems against each other can be computed. Figure 4.4 shows spectral difference errors between the acquired data. The comparison between the measurement using the reference system and the analytical solution (——) shows a small error of under 1 dB up until 8 kHz. The error is continuously increasing with the frequency and reaches approximately 2 dB at 18 kHz. This error can stem from several uncertainties in the measurement compared to the analytical solution. As the azimuth rotation of the measurement can only be controlled using the time of arrival of the impulse response, a small uncertainty concerning the correct orientation of the sphere remains. Furthermore, a mismatch between both the spheres material, and the speed of sound can potentially degrade the error without showing actual errors in the measurement system.

Both errors also apply for the comparison between the arc measurement and the analytical solution (——). This shows a slightly larger error value overall, lying between 1 dB and 3 dB between 3 kHz and 20 kHz with occasional peaks. The error is more noisy, indicating a more disturbed measurement. This might

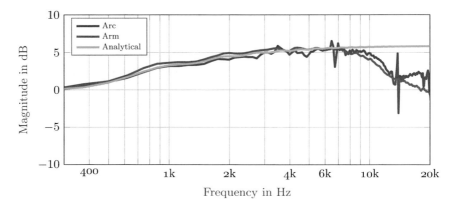

Figure 4.5: Plane wave on solid sphere at $\varphi = 0°$, $\vartheta = 0°$ shown from measurement with arm and arc and from an analytical solution.

stem from the uneven sound field, seen in Figure 4.2. A large peak in error at approximately 14 kHz certainly is the result of the resonance of the loudspeakers, already seen in Figure 3.5. A very similar error can be seen for the comparison between the two measurement systems (———). As the sphere is identical for both measurements, only a slight mismatch in orientation might remain, other than the difference in the measurement system itself. Both the noisy course and peaks can be seen in the comparison.

The error should be considered in the context of a reposition error of the fast measurement system. This error, seen in the last curve (———), is calculated from two consecutive measurements. In between the measurements, the full setup is disassembled and reassembled. This measurement is used as a second test to validate the measurement setup. A comparably low error can be seen. A critical frequency for an error increase is, as in the sound field analysis, 7 kHz. At approximately 13 kHz a resonance from the loudspeaker increases the error again.

To better compare the errors and examine the cause of the increase, single direction plots are shown in the following. Figure 4.5 shows the three data-sets for the angle $\varphi = 0°$, $\vartheta = 0°$ relative to the microphone. It corresponds to the position of the sphere, where the sound source is pointed directly towards the microphone. The analytical solution (———) shows no influence of the sphere on the sound-field at low frequencies. Starting at 300 Hz the sound pressure rises as the spheres starts to be reflective. This rise continues up until approximately

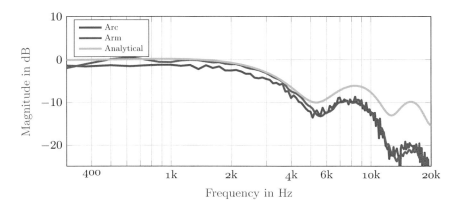

Figure 4.6: Plane wave on solid sphere at $\varphi = 160°$, $\vartheta = 0°$ shown from measurement with arm and arc and from an analytical solution.

8 kHz, where the sphere is fully reflective and behaves similar to an infinite plane. The sound pressure is increased by 6 dB. Both measurements follow the same course up until 9 kHz. Starting at this frequency, the sound pressure decreases with the frequency until 0 dB at 20 kHz. The arc measurement shows disturbances especially at frequencies 7 kHz and 15 kHz. As this decrease is present in both measurements, the cause is not disturbance from a measurement setup, but more due to the measurement conditions. In the sphere, a 1/2 " free-field microphone is used. The free field assumption is violated if the microphone is mounted inside the sphere which causes the seen deviation.

Another exemplary direction is shown in Figure 4.6. It shows the sphere measurements for the direction $\varphi = 160°$, $\vartheta = 0°$. General differences between the analytical solution and the measurements are similar to the previous direction. The measured sound pressure is lower than the calculated one. However, the difference between the solutions is bigger than in the frontal direction. One likely explanation is the influence of the gate above the microphone. This gate has a larger influence at high frequencies if the sound propagation is not perpendicular to the membrane. Both measurements are again very similar, with the arc measurement slightly more disturbed, presumably from the measurement setup.

Both shown influences of the measurement microphone create a frequency-dependent error, increasing with the frequency and causes the error shown in Figure 4.4.

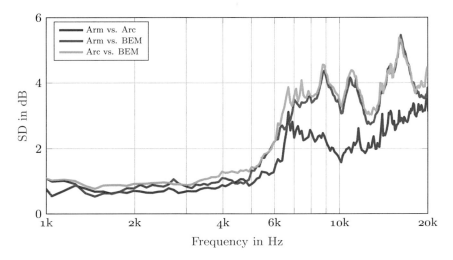

Figure 4.7: SD error of the comparison between the traditional measurement system, the fast measurement system and a BEM simulation of the left ear of an artificial head.

4.2.3 Head

Lastly, a comparison of an even more complex object is done. As the measurement setup is primarily used for the measurement of HRTF, an artificial head is a well suited measurement object. It does have all the spatial complexity and a comparable size to most human heads and can be used to test the reproducibility as well. To this end, several measurements of an artificial head are acquired in different measurement systems. As before, one measurement in a traditional measurement system, used in the same measurement distance and at the same positions, serves as a reference to compare against. Additionally, a Boundary Element Method (BEM) simulation of the artificial head is computed. While the simulation does not suffer from measurement uncertainties, such as positioning errors, influences from the speaker directivity, or reflections from the construction, other uncertainty factors influences the results, with the unknown surface impedance, or the surface of the eardrums for example. Figure 4.7 shows SD values for three different comparisons for the left ear of the head. The corresponding figure for data of the right ear can be seen in the appendix (Figure A.1). As very similar results are obtained for the right ear, in the following, only data from the left ear is shown. A comparison between the two measurement systems (———) shows a slightly higher overall value than the comparison for the simpler object shown in Figure 4.4 (———). The value is below 2 dB up until 7 kHz. At this frequency, a steep incline in error is visible which is followed by a decrease up until 10 kHz. The error increases with frequency, but remains below approximately 3 dB.

The comparisons from the measurements to the BEM simulation (——— and ———) shows an overall larger difference. Both curves are very similar to each other, which suggests that an overlaying mismatch between the datasets occludes differences from the measurement techniques.

In the following, this mismatch is examined with the use of three exemplary comparisons of single directions between the used dataset. The observations also highlight the shortcomings of the used spectral difference metric. Figures 4.8, 4.9 and 4.10 show three example directions of the left ear of the compared HRTF. Three different causes for an increased error can be seen. In Figure 4.8 a large error at 6 kHz is caused by a resonance frequency that is differently damped in all three data-sets. In a BEM simulation this can be caused by a simulation mesh that is not fine enough to sample the resonance and is a common occurrence at large resonance dips. The deviation between the measurements are likely caused by disturbances from the measurement setup itself and are of high interest for an evaluation. Both comparisons have a very large error of at least 20 dB at this frequency point. In Figure 4.9, a different error causes large errors. Here, the resonance frequencies at approximately 7 kHz, 12 kHz and 14 kHz are measured with a slight frequency mismatch between the three datasets. These mismatches might be a result of different temperatures while acquiring the datasets. A temperature shift shifts the speed of sound and thus the resonance frequency. While otherwise matching very well, the spectral differences are large at multiple adjacent frequencies. Lastly, in Figure 4.10, a mismatch between the simulation and the measurements is visible between 9 kHz and 10 kHz. In this frequency-band, the simulation exhibits two resonances that are not as pronounced in the measurements. These resonances might be the pinna material. While it is assumed to be infinitely rigid and reflective in the simulation, the real material behaves differently which could cause these resonances to be more damped.

From these observations, the need for better suited error measure becomes clear. While the undamped resonances in the simulation might be a fault that is worth noting, a perceptual model should be taken into account that assesses the relevance of these differences. The frequency mismatch of the second example is,

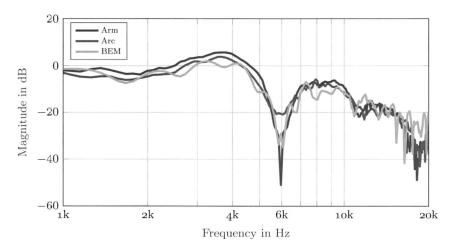

Figure 4.8: Direction $\varphi = 280°$, $\vartheta = 29°$ of the left ear of a measurement in the arm, the arc and from the BEM simulation.

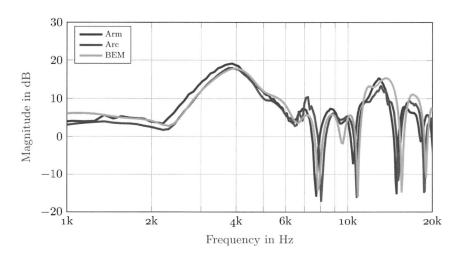

Figure 4.9: Direction $\varphi = 24°$, $\vartheta = 22°$ of the left ear of a measurement in the arm, the arc and from the BEM simulation.

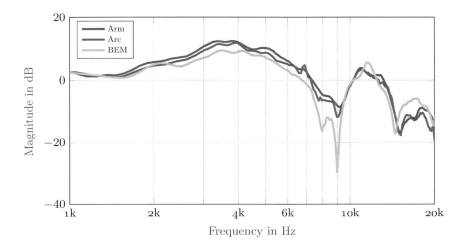

Figure 4.10: Direction $\varphi = 173°$, $\vartheta = -8°$ of the left ear of a measurement in the arm, the arc and from the BEM simulation.

most likely, of little relevance to the perception of the filter. In the future, an advanced error measure is needed which is more differentiated between audible and non-audible differences. The use of the Notch Frequency Distance metric introduced by Iida [106] combined with a just-noticeable difference for notch frequency shifts might be a good starting point to compensate for these kinds of errors.

However, as most of the discussed errors mainly affect the comparison between measurements and simulations, the error measures can still be used for comparisons between measurements which is done in the remainder of this thesis.

Additional to the comparisons between datasets to evaluate the error, uncertainties in the measurement system itself can be characterized with the repetition and reposition error. To this end, three measurements of the artificial head are acquired in the fast measurement system. Between the first and the second measurement, the artificial head is not touched. The measurements are taken subsequently with little time delay to reduce influences from time variances in the measurement room. A comparison between the two measurements therefore shows a repetition error without influences of positional errors. Between the second and the third measurement, the measurement system is disassembled and reassembled. The subsequent comparison shows the reposition error: A repetition

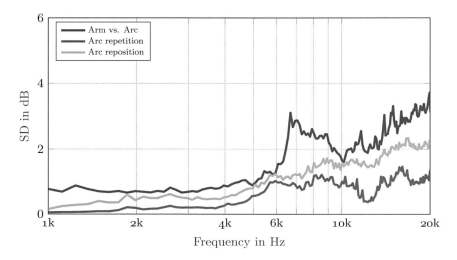

Figure 4.11: SD error of the repetition and reposition error inside the fast
HRTF measurement system.

with the additional influence of positional uncertainties. Figure 4.11 shows both
comparisons and also includes the comparison between the measurement systems
form Figure 4.7 as a reference curve. As expected, both repetition (——) and
reposition error (——) exhibit a lower overall error than the comparison between
measurement systems while the repetition error is again lower than the repetition
due to the missing influences of the position. Both errors are slightly increasing
with frequency.

Old Measurement System

For the sake of completeness, a comparison between the previous fast measurement
system, and the system introduced in Chapter 3 is done analogously. One
measurement of the fast measurement systems each are compared against a
measurement from the traditional measurement system. As the fast measurement
systems use a different measurement distance, the reference measurement is
different for both comparisons. Figure 4.12 shows the comparison. Both error
values are of comparable size and show approximately the same increase over
frequency. Overall, the errors from the previous measurement system are slightly
larger than from the new system. This increased error might by a result of

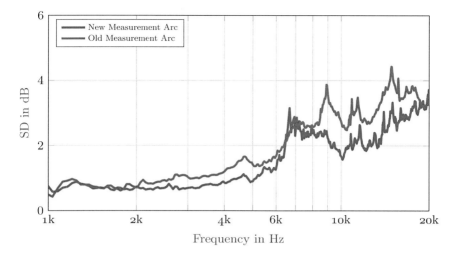

Figure 4.12: SD error of a measurement of an artificial head in both the old fast measurement system and the described measurement system, compared to a reference measurement.

reflections between the speakers, or the uneven directivity shown in Figure 3.7.

4.3 Influence of the Spherical Harmonic Decomposition

This section deals with the influence of the post-processing using the spherical harmonic decomposition, described in Section 2.6 and used in the following to interpolate HRTFs.

To evaluate the influence of the decomposition on measured data, a HRTF measurement of 5°equiangular resolution is used, decomposed to spherical harmonic coefficients, and reconstructed to the same measurement positions. A spatial difference between the original measurements and the reconstructed data gives an indication on the error introduced by the transformation. This error is caused by the order limit during the transformation, called truncation error, aliasing effects by the selected transformation order and numerical uncertainties during the inversion of the transformation matrix. These numerical uncertainties can be caused also by missing data in the sphere for example from the missing bottom cap.

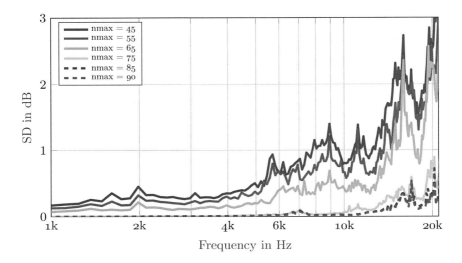

Figure 4.13: SD error for different maximum order limits during SH transformation.

Figure 4.13 shows the overall SD error for multiple upper order limits, ranging from order 45 up to order 90. It can be seen that, while the error is low in magnitude for low frequencies, with order 45 a relatively large error of 2-3 dB is present at high frequencies. Using a higher upper order limit, this error decreases as expected. For order 75 the error is below 1 dB over the whole audible frequency range. As a result of this investigation, the upper order limit is set to 75 for all transformations shown in this thesis.

4.3.1 Subjective Evaluation

To evaluate the audible difference of the spherical harmonic transformation, a listening experiment has been conducted. The test evaluates up to which order the transformation introduces audible artifacts into the measurement. To this end, a Three-Alternative Forced-Choice (3AFC) experiment design was chosen. During this test, three stimuli are presented to the subject. Two of the stimuli are taken from one random direction of the reference HRTF. The third stimulus is taken from the corresponding HRTF direction with changing upper order limit of the SH transformation. The stimuli are generated by convolving the respective selected HRTF with a pulsed noise stimulus containing three white

Table 4.1: Five selected positions for the audibility listening experiment.

φ [°]	355	70	75	80	85
θ [°]	54.2	56.7	56.7	109.6	114.7

noise pulses, each 300 ms in length with 200 ms pause in between. Two different independent variables are tested. The first variable is the upper order limit of the spherical harmonic transformation. Five different orders have been selected using a pretest. The orders have been selected as 4, 8, 12, 16 and 20. The second independent variable is the source direction. As the error varies over the sphere, an influence of the error might occur. Five directions are selected from the measured directions of the five degree azimuth resolution static reference measurement. The directions are selected in a way that the error in the direction must be monotonously decreasing between transformation order maximum of 1 to 20. The directions can be seen in Table 4.1.

To reduce the influence of guessing, each stimulus pair is tested ten times. The total number of stimuli tested for each subject is calculated as:

Directions		HRTFs		Repetitions		
5	·	5	·	10	=	250

24 subjects are tested in five randomized blocks of 50 stimuli each. The subjects have a mean age of 25.7 years with standard deviation of 4.4 years. Nine subjects are female and 15 subjects are male. Two subjects self-reported previous experience with HRTFs and binaural listening.

Figure 4.14 shows the detection rate averaged over all directions per SH order. A high detection rate signifies a high audibility in the differences. As expected, a clear decrease of audibility towards higher upper order limits can be seen. Starting at order 16, the mean of the detection rate of all subjects lies at the guessing probability, indicating that, on average, subjects are not able to tell a difference between reference and stimulus.

4.3.2 Discussion

While the subjective evaluation suggests no audible differences even at relatively low transformation orders, no general conclusion can be drawn from the test.

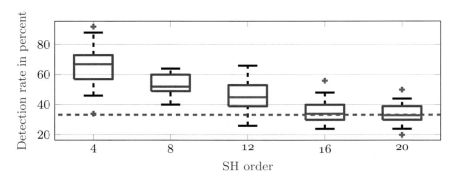

Figure 4.14: Relative detection rate over the tested reconstruction SH order, averaged over all five test directions and subjects.

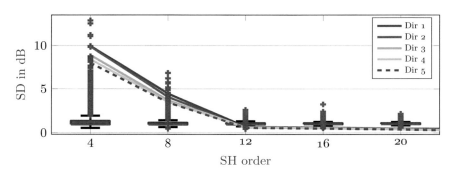

Figure 4.15: Middlebrooks spectral difference values for all directions over the SH order. All five tested directions are also shown.

This is partly down to the selection of the stimuli. Figure 4.15 shows Middlebrooks spectral differences for every direction per SH order as a boxplot. While a slight reduction of the mean error is visible, the reduction of error outliers is more apparent. The figure also shows the five selected tested directions. As can be seen, the directions all lie in the higher outliers for low SH orders and in low outliers for higher orders. This introduces a bias towards higher SH orders into the test and suggests that the average audibility of the transformation influences are different.

5

Subjective Evaluation of Localization Accuracy

One important aspect of acquiring HRTF is of course their use in binaural synthesis applications. In these applications, localization of sound sources is one factor in the quality the synthesis. This chapter presents subjective localization experiments using HRTFs from fast HRTF measurement systems. The goal of these experiments is to validate the quality of the measurement setups itself.

Three different evaluations are presented. First, experiments using static sound sources are performed. In these experiments, the stimulus consists of one single HRTF direction and is presented to the subject using headphones. If the subject moves their head during the stimulus presentation, the source stays at the same position relative to the subject's head center. Therefore the source moves in union with the head. From the three experiments presented in the following, two are experiments with static source reproduction. In the first experiment, the influence of the measurement setup itself is evaluated. In the second experiment, the gain from individual measurements is studied. The third experiment uses dynamic reproduction. Contrary to the static reproduction, the subject's head movements are tracked. From the resulting movement information, the location of the virtual sound source is updated during the movements. If the subject moves the head, the sound source remains stationary relative to the room. Generally, a reduced localization error and front-back confusion is expected with dynamic reproduction.

5.1 Localization of Static Sound Sources

This section presents the results of two separate listening experiments using static sound sources. While the static reproduction of sound sources is easy to implement and does not require additional computational steps that could potentially introduce errors, it does not represent a natural listening environment as described in Section 2.2.1. However, many previous investigations using this method exist which makes a comparison and validation of the findings easier. The results from these investigations can not be compared against results using real sources, as a presentation of a stimulus with loudspeakers would be a dynamic reproduction situation in which the subject is able to move their head to improve localization.

Because of these limitations, the two experiments are structured as follows: In the first experiment multiple directions of two measurements of the same artificial head are presented. The first measurement is taken with a traditional HRTF measurement system using only one loudspeaker. The second measurement is acquired using a fast HRTF measurement system. The comparison is done to investigate if the use of the fast system introduces so much errors that the localization ability is limited. It can be regarded as an subjective evaluation of the disturbances in the sound field found and discussed in Chapter 4. The second experiment compares the localization ability using non-individual HRTFs from the same artificial head to individual HRTFs. All measurements for this experiment are performed in the same measurement setup. This experiment is performed to gain further insight into the performance of the system. As the localization ability is suspected to be more precise with individual HRTFs, errors introduced by the system are more likely to impact the results than with non-individual HRTFs. The global performance can be compared to existing publications that make use of the same presentation and reproduction methods. This experiment further serves to quantize the quality of the acquired HRTFs as an increase in localization performance is expected. The magnitude of this increase can be an indicator on the quality. Note that both experiments use HRTF data from the previous fast measurement setup described in Section 3.1 and introduced by Masiero [96].

Firstly, the used pointing method for both experiments is introduced. This is followed by a definition of used error measures to quantize the localization error and results for both experiments. [1]

---

[1] Parts of this chapter have been published in [79].

5.1.1 Pointing Method

To evaluate the localization ability given a set of HRTF, a pointing method is required by which the subjects can indicate the perceived sound location. Different pointing strategies are introduced in Section 2.4.

The chosen pointing method uses a display in front of the subject. On this display, an arrow, representing the subject, and a sphere are shown. On the sphere a cross-air can be used to indicate a position. All elements are shown from a perspective behind the sphere looking into the same direction as the subject from a slightly elevated position. For reasons of visibility, the sphere itself is only stylized as a circle representing the horizontal plane and two quarter circles indicating the position of the frontal plane. Additionally, helper lines in five degree resolution in azimuth and elevation are plotted in a sphere segment that the cross air currently occupies. This lets the subjects perceive the three dimensionality of the sphere as intended, without occluding important parts of the sphere unnecessarily.

The cross-air itself is controlled using the analog sticks of a gamepad. This combination of display and pointing device leads to a categorization of this method as an extrinsic pointing method.

After the cross-air is set to the perceived location, the subject can push a button to fix the directional input. Alternatively, the subject is able to signal an in-head-localization when no externalization of the sound source can be perceived. In this case, the cross air vanishes and the arrow indicating the subject changes color to signify the in-head localization. Additionally, the subject has the option to flip the viewing direction to the front of the sphere using a different button to make pointing to badly visible parts of the sphere easier. The full user interface can be seen in Figure 5.1.

5.1.2 Error Measures

From the collected data, different error measures can be extracted. The most intuitive error measure is the use of differences in the indicated position described by the two spatial angles φ and ϑ. While these angles are a good way to describe points on a sphere, two different angles are better suited in the context of localization. In this context, one likely mistake is an error along a cone of

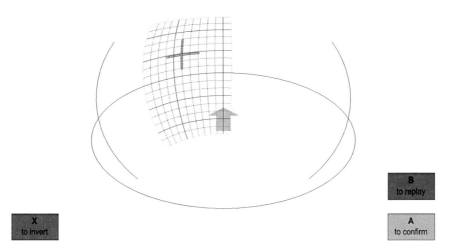

Figure 5.1: Graphical User Interface of the pointing method [79].

confusion (see Section 2.1.3) where the binaural cues are all identical. Without visual aids, small head movements, or another means to obtain more spatial information, errors along the cones of confusion are very common even with perfect individual HRTFs. An error along a cone of confusion is therefore expected and does not give much information about the quality of the binaural synthesis and the used HRTF. The quality of the presentation can be measured by deviations from the expected cone of confusion. However, both φ and ϑ change along each cone. Therefore an isolation of expected errors, meaning errors along those cones of confusions, from not expected errors, can not be achieved using these angles.

Two different angles are defined to better represent these errors along the cones of confusion. These angles are defined analog to φ and ϑ but on a rotated system. Instead of the angle from the z-axis towards the point, the angle α is defined as the angle from the y-axis towards the point and the angle β is defined as a rotation around the y-axis [107]. Figure 5.2 shows the definitions of α and β as well as the traditional angles φ and ϑ. In the context of localization accuracy, a deviation of the α angle describes a deviation from the expected cone of confusion, while the difference in β angle error indicates the error on the cone of confusion itself.

While the deviations from the desired cone of confusion can be described using the angle α, certain care has to be taken when interpreting the results of the

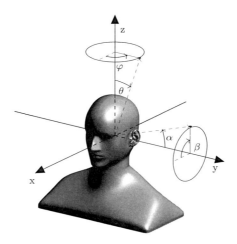

Figure 5.2: Visualization of ϑ and φ angles compared to α and β angles.

angle β. This angle artificially inflates the deviation towards the sides where the poles are [26]. The same spatial deviation at $\alpha = 5\,°$ is large in β compared to $\alpha = 90\,°$ as the cone of confusion is small in circumference near the poles.

In the following, results from the α angle and hemisphere confusion in percent are given to quantify the results. The combination of both values provides a conclusive measure of the localization ability.

Hemisphere Confusion

As an additional error measure, the so called *hemisphere confusion* is used. A hemisphere confusion is counted if the angle β is larger than $90\,°$. In these cases a confusion of the source from the front to the back, or vice versa has occurred. The sum of the occurrences of both errors relative to the total number of stimuli played is the hemisphere confusion error.

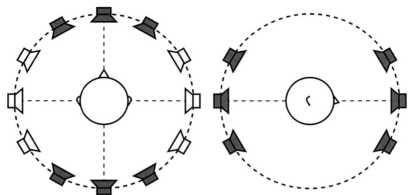

(a) Top View of all virtual source positions in the horizontal plane.

(b) Side View of all virtual source positions in the median plane.

Figure 5.3: Top and side view of all virtual sound source positions. Red speakers indicate the positions which also feature elevated positions.

5.1.3 Comparison of Measurement Setups

The first experiment deals with the question whether the fast measurement influences the localization ability of the acquired HRTF. To this end, 21 subjects are instructed to localize stimuli from measurements of an artificial head HRTF using both fast and traditional measurements. For the task, 24 different spatial directions are chosen on a sphere around the subject. Each direction is tested six times per subject to reduce bias from the pointing method, fatigue effects or learning effects. To reduce bias from different loudness of the stimuli as a result of the spatial filter, the amplitude of the finished stimuli is changed within the range of $\pm 5\,\mathrm{dB}$ in $1\,\mathrm{dB}$ steps as applied by Makous and Middlebrooks [60] and van den Bogaert [108].

From the 24 positions, twelve positions are on the horizontal plane spaced every $30\,^\circ$ starting in the frontal direction of $\phi = 0°$. The remaining twelve positions are split between positions in elevation $\vartheta = -30°$ and $\vartheta = 30°$. As shown in Figures 5.3a and 5.3b, on each elevation level there are six source positions covering the front and back directions at $\phi = -30°, \phi = 0°, \phi = 30°$ and at $\phi = 150°, \phi = 180°, \phi = 210°$ azimuth respectively. All sources are placed with $1\,\mathrm{m}$ distance to the subject, which corresponds to the measurement distance of the used fast measurement setup.

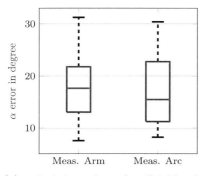
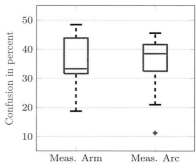

(a) α deviation values of artificial head HRTFs from both traditional and fast measurement.

(b) Hemisphere confusion in percent of artificial head HRTFs from both traditional and fast measurement.

Figure 5.4: Localization experiment results from the comparison between two different measurement setups.

The used full-phase HRTFs have a length of 512 samples. Each HRTF is convolved with a pulsed white noise stimulus with three pulses, each 300 ms long, with a pause of 200 ms. The stimuli are played back using Sennheiser HD600 headphones which are individually equalized using a method proposed by Masiero and Fels [20] (see Section 2.2.1). It should be noted that with this stimuli length no gain in horizontal localization accuracy is expected from small head movements when presenting sounds with loudspeakers [109, 110]

Results

Figure 5.4a shows boxplots of averaged results of the α deviation between presented and indicated position for both HRTF conditions. The error of all repetitions to all source position is averaged per person. It can be seen that both conditions have a similar mean error value. The mean error is approximately 18° for both conditions. The interquartile range from 25% to 75% over the two different test subject sets are also of comparable size, with a slightly bigger box and lower mean for the measurement arc. No statistical significance between the two datasets exists. Table 5.1 shows ANOVA results of the comparison of all relevant error measures. Mean and standard deviation values can be seen in Table 5.4.

For the hemisphere confusion percentages, shown in Figure 5.4b, also no statistical significant difference exists. The percentages mean range from 35% for the measurement arm to 38% for the measurement arc.

The lack of significant difference leads to the conclusion that no difference in localization ability is detected between the measurement setups. However, not that from the experiment it can not be determined, if there are no noticeable differences in spatial position, or if the pointing method introduces to much uncertainty in the position that fine differences can not be repeated.

Table 5.1: Variance analysis of in-between subject data of localization performance with HRTFs from two different measurement systems [79].

Error Measure	$F(1,36)$	p-Value	η^2
Azimuth	F < 1	$p = 0.60$	$\eta^2 = 0.01$
Elevation	F < 1	$p = 0.63$	$\eta^2 = 0.01$
α angle	F < 1	$p = 0.68$	$\eta^2 = 0.00$

The obtained results are of comparable size to previous published research. In an experiment with 9 subjects, Begault et al. [78], subjects localized speech stimuli from static sound sources with non-individual HRTFs. They report localization errors in azimuth of 21°and 18°in elevation. In this study, a large front-back confusion error of 59% is reported, while in a separate study with 11 subjects [4] the front-back error rate of 29% is reported.

5.1.4 Individual HRTFs

In a further step to evaluate the measurement setup, a second static localization experiment is conducted. To evaluate the validity of the acquired HRTF, individual measurements are compared in an identical experiment against artificial data. The expectation, founded on previous studies on localization ability using individual HRTF data, would be a significant reduction in localization error and front-back confusion errors (see Section 2.4).

In this experiment 16 subjects participated (12 male, 4 female). The subjects individual HRTF is measured at a separate date before the experiment. Measurements are acquired in $30°$ steps in azimuth and $5°$ elevation steps.

The experiment itself is identical to the previous experiment including all settings

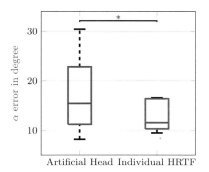
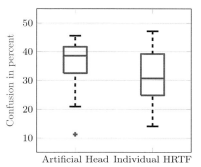

(a) α error values of HRTF condition artificial head compared to individual measurement.

(b) Hemisphere confusion in percent of artificial head HRTFs from both traditional and fast measurement.

Figure 5.5: Localization experiment results from the comparison between individual and non-individual HRTFs.

of the used HRTF, the virtual sound source positions, the number of repetition and the loudness variations to make the experiments comparable.

Figure 5.5a shows α deviation values for both artificial head data and individual measurements of 16 subjects in a within subjects comparison. As expected, the average α deviation value decreases with the use of individual HRTFs. This decrease has been found to be statistically significant. Results of the ANOVA analysis can be seen in Table 5.2.

The most pronounced reduction in localization accuracy is achieved in the elevation accuracy for elevated sound sources. Figure 5.6 shows the perceived elevation, averaged over all positions, for all three elevation levels for both subject groups using the artificial HRTF reproduction (——) and individual HRTF (——) respectively. The figure shows the increase in elevation perception clearly. For the elevated sources, the perception is noticeably closer at the sources.

Additionally, the hemisphere confusion errors for both HRTF conditions can be seen in Figure 5.5b. This error shows the percentage of a front-back or back-front confusion. A confusion is counted, if the β error is greater than 90°. As expected, a significant reduction of front-back confusion errors has been achieved by the use of individual measurements. An ANOVA analysis of the confusion errors can be seen in Table 5.3.

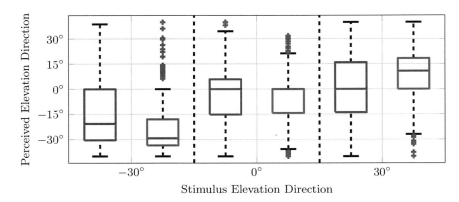

Figure 5.6: The perceived elevation averaged over all positions for all three elevation levels for both artificial HRTF (——) and individual HRTF (——) [79].

Table 5.2: Results from one-way ANOVA comparing results from artificial-head HRTFs and individual HRTFs [79].

Error Measure	$F_{(1,34)}$	p-Value	η^2
Azimuth	7.40	$p < 0.05$	$\eta^2 = 0.18$
Elevation	13.39	$p < 0.001$	$\eta^2 = 0.29$
α angle	7.44	$p < 0.05$	$\eta^2 = 0.18$

The results showed that a significant reduction in localization error could be achieved with the use of individual HRTFs. This reduction influences both the α error and the front-back confusion errors. A comparison to previously published literature shows good agreement with the results, giving credibility to the experiment. Middlebrooks [26] reports mean localization accuracy of static sources and individual HRTF of 14.5° as α angle error and 10.2° of elevation error. For this experiment, 18 subjects were used and a nose pointing mechanism was employed. In a study by Wightman and Kistler [61] eight subjects were tasked with the localization of short white noise bursts. The reported azimuth angle errors range from 16.1° to 29.8°, depending on the source direction. The reported front back reversal rate was approximately 20%.

Table 5.3: Results from one-way ANOVA comparing occurrence rates for all sector errors with artificial-head and individual HRTFs [79].

Error Measure	$F(1,34)$	p-Value	η^2
Front↔Back	6.18	$p < 0.05$	$\eta^2 = 0.16$
Up↔Down	10.98	$p < 0.05$	$\eta^2 = 0.25$
Left↔Right	$F < 1$	$p = 0.46$	$\eta^2 = 0.02$

Table 5.4: Mean and standard deviation values for α, β and hemisphere confusion error for measurement setup comparison and individual HRTF data.

		Arm	Arc	Individual
Alpha	Mean	$16.83\,^\circ$	$17.38\,^\circ$	$11.84\,^\circ$
	Std	$17.00\,^\circ$	$15.81\,^\circ$	$12.59\,^\circ$
Beta	Mean	$97.24\,^\circ$	$96.02\,^\circ$	$83.06\,^\circ$
	Std	$90.36\,^\circ$	$87.95\,^\circ$	$95.92\,^\circ$
Confusion Error	Mean	$35\,\%$	$38\,\%$	$27\,\%$
	Std	$48\,\%$	$48\,\%$	$44\,\%$
Azimuth	Mean	$19.08\,^\circ$	$18.28\,^\circ$	$12.86\,^\circ$
	Std	$17.58\,^\circ$	$18.67\,^\circ$	$14.42\,^\circ$
Elevation	Mean	$20.24\,^\circ$	$18.99\,^\circ$	$13.93\,^\circ$
	Std	$16.18\,^\circ$	$16.29\,^\circ$	$13.94\,^\circ$

5.2 Localization of Dynamic Sound Sources

An additional validation is performed using the newly constructed measurement setup described in Chapter 3. This section deals with this separate listening experiment aimed to evaluate localization accuracy when using dynamic sound sources[2].

Contrary to a static sound source used in the previous experiments, a dynamic sound source stays at the same position in a global coordinate system, even if the subject is moving. The real world analogy for this reproduction is a real speaker in a room. To achieve a dynamic source using binaural synthesis, two conditions need to be fulfilled. First, the subjects movements need to be tracked during playback. This allows to react to the movement as the relative position between source and receiver changes. Any change in angle needs to be reflected in a

---

[2]This section is based on a subset of already published data [111].

change in the used direction of the HRTF. This leads to the second requirement: A spatially dense HRTF to select the filters from.

As introduced in Section 2.4.1, humans improve the localization accuracy by using small head movements. With a dynamic reproduction, these head movements results in a change of spatial cues which is expected to enable a better localization accuracy than static reproduction as well as a reduction in front-back confusion.

5.2.1 Experiment Design

The presented experiment aims to study two different main effects. The first effect is the difference in localization accuracy from the use of individual HRTF as compared to HRTF from artificial heads. This test is comparable to the experiment presented in Section 5.1.4. The results are reported again as multiple factors in the experiment design are different which could influence the results.

The second factor is the influence of a dynamic reproduction using tracked subject movements during playback. As described in Section 2.4.1, a increase in localization accuracy is expected from the dynamic reproduction. Both factors result in four different reproduction methods listed in Table 5.5.

Table 5.5: Four reproduction methods.

HRTF	reproduction
artificial	static
artificial	dynamic
individual	static
individual	dynamic

To test both factors, eleven spatial positions are selected as defined by Table 5.6. The positions are grouped into three categories, depending on their position on the horizontal plane. The three categories are *Front*, *Side* and *Back*. The spatial position category is regarded as a third independent variable of the experiment. Each stimuli for each direction and reproduction condition is repeated four times per subject. The total number of presented stimuli can thus be calculated as:

$$\underset{11}{\text{Dir.}} \quad \cdot \quad \underset{4}{\text{HRTFs}} \quad \cdot \quad \underset{4}{\text{Rep.}} \quad = \quad 88$$

Table 5.6: Eleven spatial position tested in the described experiment.

φ	θ	Category
$0°$	$60°$	Front
$0°$	$90°$	Front
$350°$	$90°$	Front
$350°$	$120°$	Front
$300°$	$90°$	Side
$292.84°$	$70°$	Side
$247.16°$	$70°$	Side
$240°$	$90°$	Side
$190°$	$90°$	Back
$180°$	$60°$	Back
$180°$	$90°$	Back

Subjects

The test is performed by 14 right handed subjects (50% women) between 20 and 27 years (23.6 ± 2.4) old. The subjects are invited in 6 sessions. The first session is used to measure the individual HRTFs and HpTFs. In the second session, the subjects are trained using the pointing method using visual feedback. Before Sessions 3 and 4 the hearing ability of the subjects is checked using audiometry. These sessions originally were supposed to be the main experiment, but due to an error in the post-processing of the individual HRTFs, the data is not usable. The sessions can however be regarded as further training without feedback. Sessions 5 and 6 include the presented experiment with stimuli for an additional independent variable. This data is published in [111]. Each session has a duration of approximately 45 to 50 minutes to avoid fatigue of the subjects. The stimuli are presented in random order.

ITD-Matching

As it is not possible to perfectly position subjects during the individual HRTF measurement, a slight mismatch between the measured coordinates and the actual subject position in azimuth can not be ruled out. This error would also affect the reported localization accuracy, as the subjects would perceive the source at a different location than intended by the experiment setup. To correct this behavior, the reported azimuth angles are subtracted by the mismatch of the subjects ITD. For this, it is assumed that the ITD is zero at the front direction of $\varphi = 0°$. The ITD is estimated from the HRTFs using methods introduced by Katz and Noisternig [112].

5.2.2 Pointing Method

Compared to the previously presented experiments, the used pointing method is changed for the following one. The main reason for this change was difficulties with the use of the gamepad reported by subjects. This potential drawback in the accuracy, combined with a known increase in difficulty using exocentric pointing methods (see Section 2.4.1), lead to the adoption of the proximal point methods introduced by Bahu [72]. This method makes use of a hand-held, tracked pointing device that the subject positions near his head at the location where the sound is perceived. As the coordinate center for this pointing method is in the center of the head this method constitutes as an egocentric method.

As the subjects are able to retain their body and head position, closed loop localization, where the stimulus is played while the subject is pointing, is feasible using this method. As a downside, not every part of the head can be easily reached using just one arm. As a result, the selected source positions are all placed on the right hemisphere and subjects are selected to be right handed. The subjects are not limited in their movement but can behave normally.

5.2.3 Stimuli

The participants are tasked with localizing a train of pulsed white noise. The frequency range is limited between 100 Hz and 20 kHz to provide high frequency cues needed for localization tasks [113]. The pulses are composed of alternating pulses of 0.3 s and 1.2 s bursts with 50 ms on- and offset ramps interrupted with 100 ms pauses. This results in a total stimulus length of 3.7 s. This length is chosen as previous research suggested that a length smaller than 2 s does not facilitate the use of head movements sufficiently [109].

5.2.4 Results

Two main hypothesis are evaluated from the subjects responses. The first hypothesis is, analog to the experiment using only static sources, that the use of

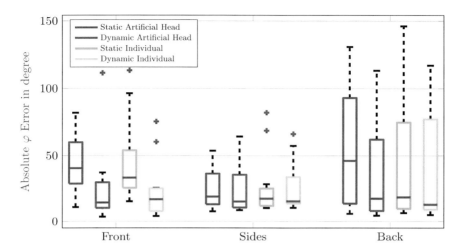

Figure 5.7: Absolute φ error for all four reproduction methods separated for the three direction groups.

individual HRTFs are beneficial to the overall localization accuracy. This hypothesis also tests the consistency and repeatability of previous experiments. The second hypothesis being an additional gain from the use of dynamic reproduction (as shown for example by Wenzel [25]).

Azimuth Error

Figure 5.7 shows the absolute error in the angle φ over the subjects. The plot shows all four reproduction methods as boxplots. Additionally, the position condition is split into the three categories *front*, *side* and *back*. The first hypothesis can be tested with the comparison of data acquired from artificial HRTFs (—— and ——) and individual HRTFs (—— and ——).

A clear reduction of the absolute error in φ can be seen for sources in front of the subject in both artificial head and individual sources when dynamic reproduction is used. At the sides and the back however, the reduction is not as pronounced. All data is corrected for front-back confusion which results in reduced values especially for front directions where the majority of confusion happens and the added angle is the largest. The presented results are therefore free of any effect that a reduction of front back confusion has and shows only the gain in accuracy.

The effect of front-back confusion is evaluated separately in the following.

The main effect, meaning averaged over both static and dynamic reproduction, as well as over all positions, does not result in a statistically significant difference, as shown by ANOVA testing. This indicates that, contrary to the expectations, no difference between artificial head and individual HRTF reproduction can be detected. However, the interaction of both HRTF and reproduction and reproduction and position reveals a significant difference. Table 5.7 shows detailed results of the ANOVA analysis.

Table 5.7: Results from ANOVA analysis of the absolute φ error values. No significant main effect, but significant interactions are found.

Absolute φ	F-Value	p-Value
HRTF	$F(1,10) = 2.5$	$p > 0.05$
HRTF x Reproduction	$F(2.54,25.42) = 3.7$	$p < 0.05$
Reproduction x Position	$F(2.23,22.27) = 3.48$	$p < 0.05$

Hemisphere Confusion Error

During localization, sources that are placed in front of the subjects are oftentimes localized in the back due to missing visual cues. Likewise, but usually not as frequently, the opposite error, a back-front confusion, can take place. Figure 5.8 shows both errors separated for all conditions and direction categories. Similar patterns as the analysis of azimuth error can be seen. In frontal directions, clear differences between static and dynamic production can be observed. This observation is in line with expectations and previous research. The additional use of individual HRTF does seem to further reduce the errors, but the reduction is not as pronounced. Significance testing using ANOVA tests revealed a significant main effect of the HRTF. Additionally, a significant main effect of the reproduction method is found. These two results confirm the expectations from previous research. The occurrence of hemisphere confusion is reduced by the use of individual HRTF and is reduced with dynamic reproduction. The main effect of the position is not significant indicating that the number of hemisphere errors does not depend on the source position. This result is contrary to the expectations, as previous research showed that the number of front-back confusions are generally higher than the number of back-front confusions. Table 5.8 shows detailed results of the ANOVA analysis.

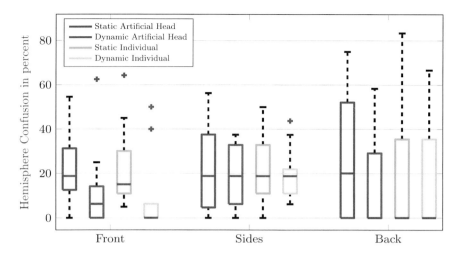

Figure 5.8: Hemisphere confusion error in percent for all four reproduction methods separated for all three direction groups.

Table 5.8: Results from ANOVA analysis of the hemisphere confusion error values. Significant main effects are found for HRTF and reproduction.

Absolute φ	F-Value	p-Value
HRTF	$F_{(1,10)} = 8.69$	$p < 0.05$
Reproduction	$F_{(1.38,13.72)} = 13.92$	$p < 0.05$
Position	$F_{(1.37,\ 13.73)} = 1.73$	$p > 0.05$

Alpha Error

A combination of both errors can again be seen in the α error values. These values are independent from hemisphere confusion and are thus well suited for an objective evaluation. Figure 5.9 shows the α error of the two reproduction methods with both non-individual and individual HRTFs. For this error measure, the main effect of the reproduction is also significant ($F_{(3,36)} = 9.403$, p < 0.05). A Bonferoni post-hoc test showed significant differences between the static artificial head reproduction and both dynamic artificial and individual HRTF reproduction methods. Mean difference values, standard error and lower and upper confidence intervals are shown in Table 5.9. This result reflects the expectation so far that there is significant difference in using dynamic reproduction when using artificial heads or individual HRTF. However, there is no difference in

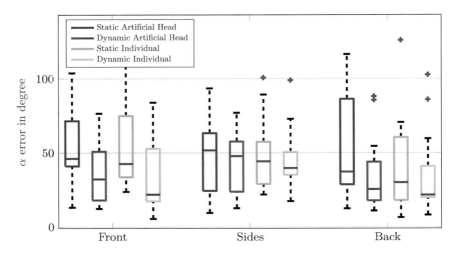

Figure 5.9: Absolute α error for all four reproduction methods separated for the three direction groups.

Table 5.9: Results from Bonferoni post-hoc test on different reproduction methods. Only data significant compared to static, artificial-head reproduction is shown.

Error Measure	Mean Diff.	Standard Error	Lower CI	Upper CI
Dummy Dynamic	10.89	2.27	3.75	18.04
Ind. Dynamic	10.40	2.42	2.78	18.03

the static reproduction between the methods which is inconsistent with previous experiments. No other main effect is found. The influence of the stimuli direction is not significant ($F(1.14, 13.677) = 1.953$, $p > 0.05$) after Greenhouse-Geisser correction for sphericity. Furthermore, no significant interaction could be found.

5.2.5 Conclusion

In this chapter, the results of three different localization experiments were presented. The goal of each experiment was to study a different aspect of the measurement setup. The first two experiments investigated static sound source reproduction, while the third experiment used dynamic sound sources. The first experiment compared HRTF from a traditional measurement against a fast measurement. No significant differences could be detected indicating either no audible difference between the datasets or differences too small to be shown by the used pointing method. The second experiment investigated localization accuracy using individual HRTF. Here, a significant reduction in both localization error and front-back confusion could be found in line with the literature. This indicates that the quality of the acquired HRTFs are performing as expected and no quality limitations could be found.

As the HRTFs used in the first two experiments are acquired using a previously used fast measurement setup, the experiment is repeated using the setup described in Chapter 3. Additionally, the experiment tests the influence of a dynamic reproduction. The results show expected trends. For α error values, a significant difference between static and dynamic reproduction is found. Additionally, a significant reduction of hemisphere confusion was found. This is in accordance to findings of Begault and colleagues who measured confusion rate differences up to 31 % [114]. For the difference between non-individual and individual HRTFs a significant reduction of hemisphere confusion of about 3 % was found.

However, no significant difference could be found when comparing α error values for static artificial head reproduction to individual HRTF which is not according to expectations. An imbalance in the chosen measurement positions might be a cause of this as the differences for lateral sources are small.

The overall results from this chapter indicate that the performance of the measurement system is in good agreement with expectations based on localization tasks from literature. However, several conclusions can be drawn with regard to an evaluation of the quality of the acquired HRTFs. From experiment one, the conclusion is drawn that the measurement system does not add so much error as to falsify localization cues in the HRTF. Experiment two and three show that localization using individual HRTF from the fast system works with comparable accuracy than systems used in the literature. The results of this investigation further confirm that dynamic reproduction enables a more accurate localization compared to static reproduction.

6

Continuous Rotation

As described in Section 3.4, two different measurement methods are available for the HRTF measurement. This section evaluates the influence of continuous rotation during measurement. Only influence of the measurement procedure itself is discussed while the influence of the measurement on involuntary subject movement is detailed in Chapter 7.

Firstly, the expected time advantages of the continuous motion is shown. A detailed analysis of multiple measurement uncertainties is followed by an objective evaluation of different rotational speeds. The additional measurement uncertainties' influence stems from multiple different sources. The biggest influences are caused firstly by the Doppler-Effect which, as the subject or the receiver is moving relatively to the sources shifts all frequencies of the measurement sweep and secondly, the changes in measurement positions even dependent on the frequency during the measurement.

The goal of this chapter is to find a continuous rotation speed that reduces the measurement time, while still resulting in an HRTF measurement that is not audibly different from a step-wise measurement while retaining the measurement signal and, to an extend, the post-processing [1].

6.1 Measurement Duration

Firstly, an analytical examination of the expected reduction in measurement duration is shown. It should be noted that only the differences caused by the rotation itself are investigated. Differences that can be achieved by, for

---

[1] The contents of this chapter have been previously published [115].

example, increasing the sweep-rate are not discussed as these steps would also be detrimental to the SNR of the measurement itself.

The duration of a full step-wise measurement is given by Equation (3.2) as

$$t_{\text{full,stepwise}} = t_{\text{single}} + (N - 1)\left(t_{\text{single}} + t_{\text{reposition}}\right).$$

This signal is constructed as a single sweep of length t_{single} after which the system moves to a new position which takes $t_{\text{reposition}}$. After the reposition, a new single sweep is played. This process is repeated for N measurement positions.

The continuous measurement sweep can be constructed analogously to the sweep t_{single}. The loudspeakers start their sweeps consecutively delayed by t_{wait}. After the last loudspeaker was started, the first loudspeaker is again started, delayed with t_{wait}. This way, each loudspeaker plays a defined number of repetitions (N), until the rotation is completed. Only at the very end, the stop-margin is added to allow for the last measurement to decay. The full measurement duration can be calculated according to Equation (6.1):

$$t_{\text{full,cont}} = (NL - 1)\, t_{\text{wait}} + t_{\text{sweep}} + t_{\text{st}}. \tag{6.1}$$

The difference between Equation (3.2) and Equation (6.1) is shown in Equation (6.2):

$$t_{\text{difference}} = t_{\text{full,stepwise}} - t_{\text{full,cont}}$$

$$t_{\text{difference}} = (N - 1)\left(t_{\text{sweep}} + t_{\text{st}} + t_{\text{reposition}} - t_{\text{wait}}\right). \tag{6.2}$$

From this Equation, a clear reduction in measurement duration can be seen. The reposition time is of course completely eliminated. Furthermore, the stop margin could be reduced to just one occurrence at the very end. Lastly, as the measurement procedure does not have to wait for the last loudspeaker to finish the sweep, but instead the measurement signal continuous, more time is saved.

6.2 Doppler-Shift

The firstly discussed uncertainty results from the rotation itself. If any sound source moves relative to the receiver, the emitted sound waves are frequency shifted by the Doppler-Effect. The frequency shift differs depending if the subject is moving relatively to the source, or vice versa. In the following the magnitude

of the shift is calculated for a moving receiver. The ratio between the emitted
and received frequency can be calculated as

$$\frac{f_{\text{shifted}}}{f} = 1 + \frac{v}{c}, \tag{6.3}$$

with c as the speed of sound, v as the receiver speed towards or moving away
from the source, f as the emitted and f_{shifted} as the received frequency. The
maximum speed, depending on the position of the source on the circle around
the subject can be calculated as

$$v_{\text{max}} = r \cdot \frac{2\pi}{T}, \tag{6.4}$$

with T as the time needed for a full rotation and r as the radius of the head (half
the distance between the ears). From the frequency ratio, a shift in cent can be
calculated as

$$x = 1200 \cdot \log_2 \left(\frac{f_{\text{shifted}}}{f} \right). \tag{6.5}$$

To calculate a sensible upper boundary for the shift, a rotational speed of $15\frac{°}{s}$
and a head radius $r = 10\,\text{cm}$ is assumed. For these values the deviation in cent is
calculated to $x = \pm 0.1317\,\text{Cent}$. The human perception of frequency differences
is frequency-dependent and has been subject to many investigations [116]. At
the most sensitive frequency range the JND value is given at around 3 cent
[117]. This leads to the conclusion that the Doppler effect is negligible for this
measurement case, as expected shifts are substantially lower than noticeable.

6.3 Measurement Positions

The changes in measurement positions resulting from the continuous rotation
have the largest influence on the quality of the measurement.

As the sweep rate of the sweeps is fixed from optimization [47], a faster rotation
means that less full sweep measurements are performed. Consequently, a faster
measurement rotation has a less densely sampled HRTF. Starting from the
measurement points an interpolation to a higher spatial density can be performed.
The influence of these interpolation methods have been subject to multiple
investigations as described in Section 2.5. The largest influence of the rotation
is the change in measurement positions. Figure 6.1 shows the influence of the

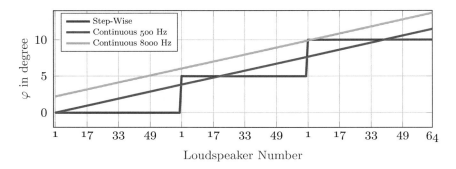

Figure 6.1: Influence of the measurement speed on the relative φ position. For each loudspeaker the figure shows φ angles for a step-wise measurement, and a continuous measurement for 500 Hz and 20 kHz

rotation on the relative azimuth angle between the speakers and the source for three full sweeps of all 64 loudspeakers. It shows a step-wise measurement (———), where the azimuth angle is constant for all loudspeakers and moves to the next position only after the measurement at the current position is finished.

Compared to this, Figure 6.1 also shows the second impact of the rotation. With the continuous rotation, the subject moves during the playback of the measurement sweeps. This impacts the measurement, as higher frequencies are measured at a different angle compared to lower ones. The figure shows two exemplary φ angles for 500 Hz (———) and 8 kHz (———) respectively.

Both described changes in measurement positions need to be corrected during post-processing of the HRTF to obtain a valid data-set. The offset between the loudspeakers can be calculated with the use of the rotational speed v and the time delay between the sweep starts of the loudspeakers t_{wait}. Equation (6.6) shows this calculation with l for each loudspeaker.

$$\Delta\varphi(l) = v \cdot t_{\text{wait}} \cdot (l - 1) \tag{6.6}$$

For the frequency-dependent shift, the relation between time and frequency of the sine sweep is needed. This can be obtained by inverting the calculation formula of the sweep rate [118]

$$t(f) = \log_e \frac{f}{f_0} \frac{\log_2(e)}{r_s}, \tag{6.7}$$

with f_0 as the start frequency of the sweep and r_s as the used sweep rate. From

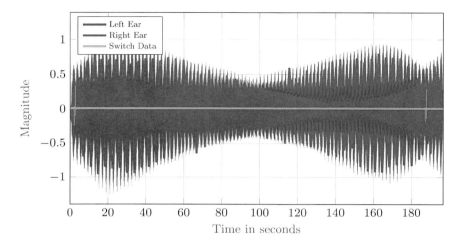

Figure 6.2: Raw, uncut, measurement data of the artificial head with more than 360° rotation. The raw measurement of all sweeps for the left and right ear are shown. The data of the position switch is measured at the same time to synchronize the position to the measurement.

both Equation (6.6) and (6.7) a total relative offset can be determined.

As a correction, a frequency-dependent spherical harmonic transformation is applied. This transformation, as described in Section 2.6, takes the frequency-dependent sampling, resulting from the rotation, and interpolates to one global, frequency constant sampling.

To determine the exact rotational speed and to establish a connection between the raw measurement signal and the position of the subject, multiple methods are possible. One is the use of a tracking device that returns the exact position and orientation of tracking bodies. However, the acquired tracking data need to be be exactly assigned to each measurement which is not easily done. For the measurements presented in this thesis, a different technique is used. A hardware switch is connected to the motor of the arc or the turntable. This switch defines one fixed position on the 360° measurement turn. When the switch is pressed, a voltage signal from the switch can be measured. This signal is directly connected to the sound-card as a third measurement signal. The measurement itself covers more than 360° to activate this switch two times, one time at the beginning of the measurement, and one time at the end. Figure 6.2 shows this exemplarily for one measurement. In the third channel, two peaks from the switch can be seen.

Table 6.1: Speed parameters of all continuous measurements. Bold values are referred to in the following.

Time [min]	0.5	1.0	1.6	2.0	3.3
Speed [°/s]	**11.2**	**5.7**	**3.8**	**2.9**	**1.8**
Repetitions	16	32	48	64	104
Resolution [φ in °]	22.5	11.2	7.5	5.6	3.4

With the assumption of a constant velocity, the overall rotational velocity can be calculated from the time differences between the peaks.

6.4 Objective Evaluation

To evaluate the error made by continuous rotation, an objective comparison is shown in the following. For this comparison, five measurements of the artificial head rotated with different rotational speeds have been acquired. They are compared against a reference measurement of the same head in the same measurement setup. The reference measurement is acquired using a step-wise measurement. In between the measurements, measurement setup remains unchanged and the artificial head is not moved in order to reduce the influence of positioning errors. As a comparison, error measures described in Section 4.1 are used. Table 6.1 shows defining parameters of all five measurements, with the measurement duration, resulting angular resolution and number of full sweeps. The parameter by which the measurements are further referred to is the rotational speed in °/s. The five measurements range from 11.2 °/s for the fastest measurement to 1.8 °/s for the slowest.

Firstly, the influence of the position correction is shown. To this end, post-processing with and without applied corrections are compared against the reference measurement. The influence is mainly visible in the phase of the difference. Figure 6.3 shows this difference for three selected measurements. It can be seen that with increasing frequency, the deviation increases for all measurements. This trend fits expectations. The rotation has an increased influence with higher frequencies. Also visible is a clear reduction of deviation with the introduced position correction compared to cases without correction. This improvement is most visible with the measurement using faster rotation. Here, the deviation can be reduced from approximately 75 ° deviation at 20 kHz to approximately 35 ° at 2 kHz making it comparable in error to a measurement, which takes 60 % longer.

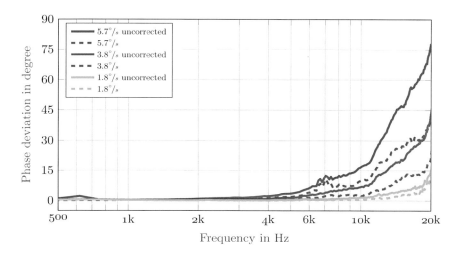

Figure 6.3: Influence of the position correction. SD phase values for a selection of three different speeds are shown relative to the reference with and without correction.

Figure 6.4 shows spectral differences in dB between all continuous measurements and the reference measurement. Two trends are visible in this. Firstly, a trend to higher error values for higher frequencies can be seen. This trend is again expected. The second visible trend is the reduction of error towards lower rotational speeds. While the error is at approximately 4 dB for measurements with fastest rotation at 6 kHz and above, the difference is reduced to lower than 2 dB for most of the frequency range. Similar results can be seen for phase deviations depicted in Figure 6.5. While the phase error exceeds 90° for a rotational speed of 11.2 °/s, it is reduced below 10° for the slow measurement with a speed of 1.8 °/s.

Very similar results can be seen in the comparison of interaural spectral differences described in Equation (4.5). Overall, Figure 6.6 shows the same errors and trends as the monaural comparisons. This observation is interesting nonetheless as it shows that both error measures are comparable and give the same approximation for the spectral differences.

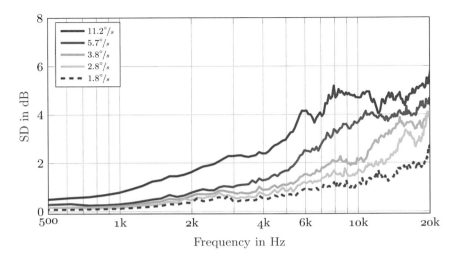

Figure 6.4: Influence of the measurement speed. SD values for five different measurement speeds are shown relative to the reference (with position correction).

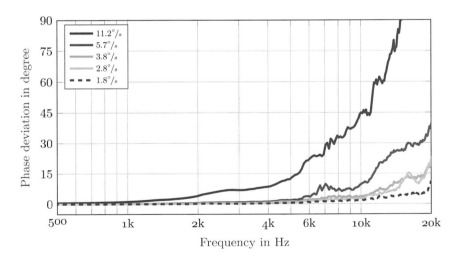

Figure 6.5: Influence of the measurement speed. Phase differences for five different measurement speeds are shown relative to the reference (with position correction).

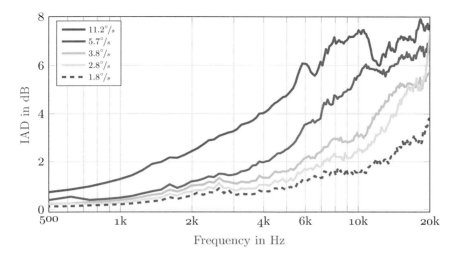

Figure 6.6: Interaural differences between all continuous measurements and the reference measurement.

6.5 Subjective Evaluation

To evaluate the audible influence of the continuous rotation, a listening experiment was performed. The goal of the experiment was to, on the one hand, validate the objective evaluation and show that no audible difference are present when comparing continuously measured and step-wise measured HRTF. On the other hand, the evaluation is done to determine an upper limit of the rotational speed at which the differences become audible.

6.5.1 Experiment Design

The experiment was designed as an 3-AFC comparison test. The subjects are presented with three random stimuli, two of which are taken from one direction of the reference measurement. The third stimulus is selected as the corresponding direction of one of the continuously measured HRTFs. The task of the subjects is to identify one stimulus that sounds different either in coloration or direction. The evaluation looks into the detection rate of each stimulus. The higher the rate, the more likely do the detections result from audible differences and not of

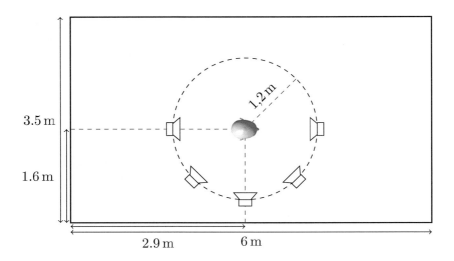

Figure 6.7: Test positions. Five positions on the median plane. Simulations are done with and without the room.

random guesses.

The first variable that is tested is the rotational speed. The five different rotational speeds that where measured are tested each against the reference measurement. From each of these data-sets, five different directions are chosen as the second variable. As the HRTF is highly directional-dependent, the selection of the stimulus has a large influence as the error distribution is not equal on the sphere. To make the results comparable to similar research, five typical directions are selected. The directions are located on the horizontal plane on the right hemisphere from $\varphi = 0°$ to $-180°$ azimuth in $-45°$ steps. The test positions can be seen in Figure 6.7. As the limited number of tested directions do not use a full spherical HRTF and thus do not represent the full spherical error, a third independent variable is introduced. For this part of the experiment, the same source positions are used and are additionally auralized inside a simulated room. The simulation results in a Binaural Room Impulse Response (BRIR) that is used for the experiment.

Figure 6.7 also shows the size of the room and the receiver position inside it. The dimensions of the room are $6\,\mathrm{m} \cdot 3.5\,\mathrm{m} \cdot 2.4\,\mathrm{m}$. The listener position in the room is located at $(2.9\,\mathrm{m}, 1.6\,\mathrm{m}, 1.3\,\mathrm{m})$ and is thus slightly off-center in all dimensions.

For the sake of repeatability, the design of the room was not focused on a realistic representation of a real room. All walls are set with the same material. Its reflection factor is set to $R = 0.4$ for all frequencies and it also has a frequency-independent scattering value of 0.3.

The simulations are calculated using the software Raven, developed at the Institute for Technical Acoustics at RWTH Aachen University [119]. The software is based on geometric acoustic simulation and uses a hybrid form of image source and ray tracing algorithms.

To increase the experiment's accuracy, each stimulus for each direction-, HRTF- and room-condition is repeated a total of six times in a random, latin-squared, fashion. The total number of stimuli per subject can therefore be calculated as

$$ \underset{5}{\text{Dir.}} \quad \cdot \quad \underset{5}{\text{HRTFs}} \quad \cdot \quad \underset{6}{\text{Rep.}} \quad \cdot \quad \underset{2}{\text{Room}} \quad = \quad 350 $$

Either HRTF or BRIR are convolved with a raw mono stimulus for playback. This stimulus is a pulsed white noise stimulus of 650 ms length. Two pulses of 300 ms length with a pause of 50 ms are used. The stimulus is band-limited between 1500 Hz and 15000 Hz. The lower limitation was used as the error variations at lower frequencies were very low and the HRTF data was obtained by interpolation, rather than measurement. The higher limitation was chosen as hearing threshold of the tested age group are already severely affected for higher frequencies [120].

A total of 21 subjects, 18 male and 3 female, participated in the experiment. They are 25.6 years of mean age with a standard deviation of 3.1 years. 15 subjects self-reported experience using listening with HRTF.

6.5.2 Experiment Results

Figure 6.8 shows results of the listening experiment described above. The boxplots show the detection rate of all subjects, averaged over all directions and plotted per HRTF measurement. Only results for the condition without room are shown. The measurement with fastest rotation shows the biggest detection rate. This signifies the largest audibility and is according to expectations. With decreasing rotation speed, the audibility is likewise decreasing. To test the results against

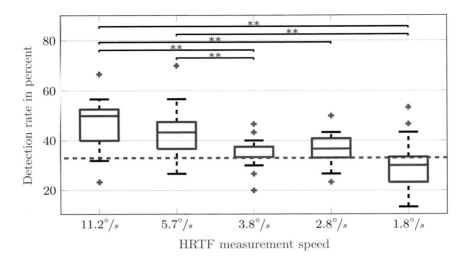

Figure 6.8: Detection rate of all subjects over all directions plotted per HRTF measurement. Stimulus without room simulation.

the guessing probability, a simulation of 10000 subjects who are always guessing is done. A comparison between this simulation and the listening experiment results shows only a significant difference for the fastest two measurements. This leads to the conclusion that the slowest three measurements do not provide any audible differences. This conclusion is also backed by an ANOVA test between the measurement speeds. Significant differences in decline in audibility can be found between the fastest, the third and the slowest measurement.

Figure 6.9 shows the same results for both room conditions. The detection rates for stimuli with room simulation are lower overall. Because of this, the decline in audibility towards lower rotational speeds is not as pronounced. This observation is in contrast to expectations as the reasoning behind adding a room simulation was to increase the number of tested directions and thus increasing audibility of differences. The difference in the room-condition is found to be significant.

The second independent variable, the source direction, is used to show this decline in detection rate. Figure 6.10 shows the detection rate over the source direction averaged over all HRTFs. It can be seen that for a presentation without room simulation, the detection rates increase for lateral directions. The difference between the room conditions is significant in the main effect ($F(1,20) = 11.03$,

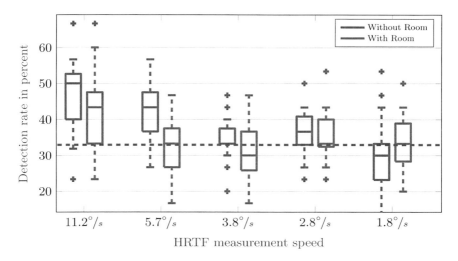

Figure 6.9: Detection rate of all subjects over all directions plotted per HRTF measurement. Comparison between the two room conditions.

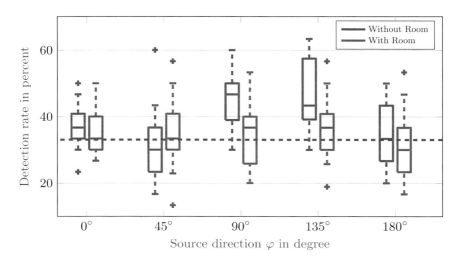

Figure 6.10: Detection rate of all subjects over all HRTF measurements plotted per directions.

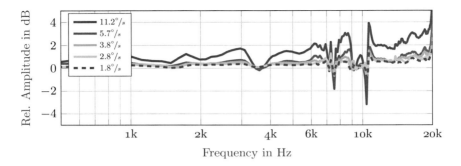

Figure 6.11: Deviation between reference measurement and all tested HRTF measurement speeds for source direction $\varphi = 0$ for the left ear.

p < 0.05) as found by an ANOVA test. In the front and back of the subject the detection rates are low. This phenomenon can not be observed in cases with room simulation. This can be explained with the use of Figures 6.11 and 6.12. The figures show differences for the left ear for only the directions at $\varphi = 0°$ and $\varphi = 90°$, respectively. While the errors for the frontal source direction are overall very low, very large deviations can be observed for counter-lateral source directions. These large deviations can be attributed to the measurement. As HRTFs have a larger number of resonance dips and notches at the counter-lateral side, the quality of the spatial reproduction is dependent on the spatial sampling. For higher rotational speeds, the spatial sampling is less dense, making a representation prone to error, even after interpolation. A similar directional dependency was shown in papers related to interpolation of HRTF before [84]. The large errors seen in Figure 6.12 have been found to be less audible in the presentation of stimuli with room simulation. This phenomenon can be explained with the use of the diffuse field HRTF described in Section 2.2. Figure 6.13 shows differences in diffuse field HRTF for all measurements. The errors are small for all measurements and frequencies. Only for the fastest measurement the deviations exceed 2 dB at frequencies above approximately 7 kHz. This small difference results in very similar sounding late reverberation of the stimuli when sound comes in a diffuse manner from all directions at random. This reverberation tail seems to be masking the differences in the direct sound.

The question of the fastest, non audible measurement speed is therefore dependent on the scenario in which the acquired HRTF is intended to be used, as more errors are masked when sound is auralized in a simulated room. For a pure direct sound auralization, the required speed for a HRTF dataset with 5° spatial

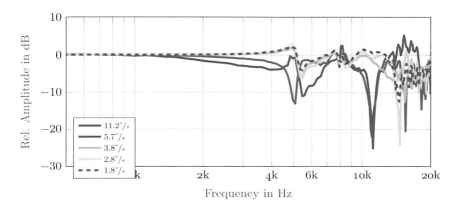

Figure 6.12: Deviation between reference measurements and all tested HRTF measurement speeds for source direction $\varphi = 90$ for the left ear.

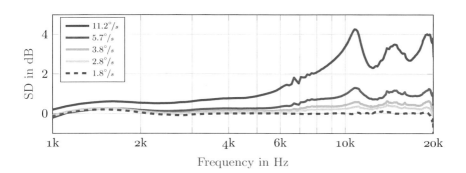

Figure 6.13: Differences in diffuse field HRTFs compared to reference measurement. Left ear only.

resolution could be reduced from seven minutes for the step-wise measurement to approximately two minutes using continuous rotation without audible artifacts in the resulting dataset.

6.6 Summary

In this chapter a objective and subjective evaluation of a continuous rotation measurement method has been presented. This method is used to further reduce the measurement time as well as reduce subject movement during the measurement. The evaluation compares the continuous measurements to a step-wise approach, where subjects are stopped at one angle for the measurement.

Six high density measurements of an artificial head have been performed. The measurements include one reference measurement with discrete azimuth resolution and five measurements with continuous azimuth rotation and increasing speed.

The compensation algorithm is evaluated using objective comparisons of the phase deviations. A comparison of the continuous measurement against the reference measurement confirmed expectations regarding a trend towards lower deviations for lower rotation speed. For the measurement with slowest rotation, the error lies below 2 dB until approximately 15 kHz.

Furthermore, a subjective evaluation is presented to investigate the audibility of the faster measurements. To this end, an 3AFC comparison against the reference measurement is performed to validate findings and to find an upper limit at which the rotation becomes audible.

However, the audibility of the introduced distortions were low. Only the fastest two measurement speeds could be differentiated significantly from the guessing rate. Additionally is was shown that the differences could be further masked through a room simulation, as the measurements did not vary much for diffuse sound fields

Ultimately, it was shown that the measurement time for a highly dense spherical individual HRTF measurement could be reduced from approximately seven minutes using a step-wise measurement to approximately two minutes for 4096 positions (5° azimuth resolution) without any audible differences.

7

Subject Movement

One important uncertainty during the measurement of individual HRTFs is the movement of subjects during the measurement. To accomplish a perfectly undisturbed measurement, the subject needs to be perfectly motionless. This need has been identified even during the early individual measurements. Searle et al. fix subjects during the measurements using a dental bite plate [38]. They give an estimated $1°$ to $2°$ error in reproducibility for the subjects position. Wightman and Kistler use a similar approach [41] for a study requiring approximately 60 minutes of measurement time. A less intrusive approach is used by Algazi [121] where a chin rest is used to control subject positions during measurements. With advances in computer technology, several research groups use head tracking systems to monitor subject movements. Feedback systems are used in a study by Møller and Sørensen [24] allowing subjects to correct their positioning themselves. If movement during a measurement occurs, the measurement result is discarded. A very similar approach is used by Bronkhorst [27], who pre defines a zone of 0.75 cm in position and $5°$ in orientation, in which a measurement is accepted.

In most studies, with the exception of the study by He [122], head movements are considered detrimental. He encourages the subjects to freely move around one loudspeaker which plays a continuous measurement signal. From the relative position between subject and loudspeaker, the corresponding HRTFs are calculated using a Normalized Least Mean-Square approach.

While the above mentioned studies try to eliminate movement, the question of the influence of these movements are not considered. This influence is studied by Riederer [123] who compares the influence of head tilting and pivoting between multiple measurements. He concludes that "pivoted head position [...] seems to cause the strongest / most wideranging alteration [...]".

The total amount of movements during a measurement also has been of interest in multiple studies. Hirahara et. al. [124] measured HRTFs over 95 minutes on three sitting subjects. No headrest was provided to stabilize the head position during the measurements. Differences in head orientation between start and end of the measurement are reported as "1° in roll but as much as 10° in the pitch and yaw directions". In a substantially bigger study, Carpentier et. al. [125] measured HRTFs of 54 subjects with head tracking, but fails to give any value on the actual head movement. A live feedback system is also employed by Denk et. al. [126]. The system displays the offset of the subject position and orientation to the subject who can adjust and correct its position during the measurement. With the use of this system, the head orientation is stabilized during the 90 minute measurement to values within 0.5°.

This chapter provides an evaluation of both the amount of subject-movement depending on the setup of the measurement setup and the influence of the movement on the measurement itself. To this end, the movements of 16 subjects during four different three minute full spherical continuous HRTF measurements are evaluated. The subjects are either sitting, standing, are either rotated or have the measurement setup rotated around them, or have a visual feedback, comparable to the work of Denk or Møller. To evaluate the measurement error, five measurements of one artificial head are acquired. During the measurements, the head is rotated by different amounts with the use of a turntable. These distorted measurements are compared to a reference measurement. Additionally, a correction of the movement with the use of head tracking is presented and evaluated. This correction is an extension of the correction presented in Chapter 6.

In the following, head orientation is given in three rotation angles *roll*, as a rotation along the frontal plane, *pitch*, as a rotation along the median plane, and *yaw* as a rotation along the horizontal plane. Figure 7.1 illustrates the three rotation angles analog to Figure 2.1[1].

7.1 Amount of Movement

As a first step, the amount of movement and how it can be influenced by the measurement setup are of interest. Zillekens [128] showed in some preliminary tests at the Institute of Technical Acoustics (ITA), that a simple neck rest reduces

---

[1]Parts of this chapter have previously been presented [127].

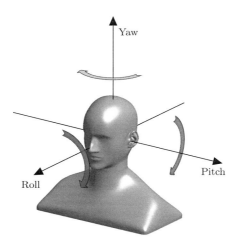

Figure 7.1: Illustration of the three commonly used rotation angles *roll, pitch* and *yaw*.

position errors of the subjects significantly, compared to standing unaided. He also showed that an additional back-rest does not provide any further benefits for the subjects.

With these findings in mind, a study of the amount of subject's movements, depending on the measurement system is conducted. In total 16 subjects are tested in four different measurement system configurations. Three different parameters are changed between the conditions. The first parameter is the subject position, with the subject either standing or sitting in the center of the measurement arc. The second parameter concerns the rotation. The rotation can either be the subject rotating around its axis with the measurement arc remaining stationary, or the measurement arc rotating around the still subject. For the third parameter an additional graphical user interface, inspired by Denk et al. [126], is used to help reduce the movement. Each parameter is changed in one condition compared to a default measurement condition with a standing subject, device rotation and without display. All four conditions are shown in Table 7.1. To compare the subject movement in both position and orientation, the raw tracking data of the case turntable needs to be corrected. In the raw data the turntable rotation is superimposed to the subject data. This correction

Table 7.1: The four measurement conditions.

Condition	Position	Rotation	Display
Default	Standing	Device	Without
Turntable	Standing	**Subject**	Without
Chair	**Sitting**	Device	Without
Display	Standing	Device	**With**

is shown in the following before comparing the results.

7.1.1 Turntable Correction

The additional rotation of the subject influences several degrees of freedom. The orientation is only influenced in the *yaw* angle. Assuming a coordinate system centered in the rotational center of the turntable, both the subject's and the turntable's tracking position can be described in cylindrical coordinates. While the angle φ of the turntable increases linearly over time (disregarding uncertainties in its movement), the difference in φ angle between the turntable and the subject is caused by the subject's movement.

To extract the *yaw* angle from the superimposed tracking data, a simple subtraction is sufficient:

$$\gamma_{\text{subject,tracked}} = \gamma_{\text{subject}} - \gamma_{\text{turntable}}, \qquad (7.1)$$

with γ representing *yaw* angles.

In the subject's position, two degrees of freedom need to be corrected from the subjects rotation. As the subject's position generally does not vary as much, the correction is not detailed here. A full description of the correction can be found in the work by Wepner [129].

7.1.2 Head Calibration

The subject's movement is observed with an optical tracking system. A tracking body, fitted to a hair-band is placed on top of the subject's head. As each subject uses the hair-band slightly differently, a calibration of the tracked hair-band to the actual center of the head is necessary. Additionally, this calibration corrects

(a) Subjects tracking hair-band.   **(b)** Head calibration pin.

Figure 7.2: Head tracking and calibration equipment [129].

all position and orientation values to be head-centric instead of centered around the tracking body on top of the head.

The calibration works by marking both ears and the base of the nose with a specialized calibration pointing pen. All points are then used relatively to the hair-band that has to remain stationary on the head during the calibration and the subsequent measurement. Figure 7.2a shows the hair-band with tracking body, while Figure 7.2b shows one calibration pin.

The head calibration pins are of well defined length, so the position of the point without tracking bodies can be calculated from the tracked position. The two ear markers are used to approximate the head center. From these positions, the head center is calculated as the halfway point between the ears according to Equation (7.2):

$$P_{head} = 0.5 \cdot (L + R),\qquad(7.2)$$

with P_{head} as the sought middle position of the head and L and R as the positions of the left and right ear. A general offset from the head position can then be calculated as

$$\vec{e}_{offset} = T - P_{head},\qquad(7.3)$$

with T as the tracked position of the hair-band. Figure 7.3 shows the relationship between the tracked points and the resulting head center point.

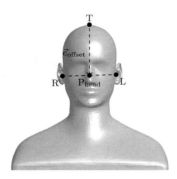

Figure 7.3: Schematic depiction of the head calibration. T as the tracked hair-band and R and L as the right and left ear respectively. Points R and L are recorded relative to T.

7.1.3 Feedback Display

As in the previous studies by Møller and Sørensen [24], Bronkhorst [27] or Denk [126], a live feedback system is used to let the subjects control their own position and orientation throughout the measurement in test case *display* (see Table 7.1) .

The goal of the design of the display is to be as easy understandable for the subject as possible. With this prerequisite, six degrees of freedom are displayed in a 2D space in an easy to understand fashion. The interface shows an unmoving cross air indicating the desired position. Additionally, the subject's real position is plotted in a second cross air. From the position of the cross air alone, deviations in two positional degrees of freedom are easily recognizable: The subject's up-down position direction, and left-right position movements. A *roll* movement can also be visualized easily.

To cover the remaining three degrees of freedom, front-back position changes as well as yaw and pitch orientation changes, additional information are added to the interface. The subject's front-back movement is visualized using the radius of the cross air. If the subject is too close to the display, the radius is increased and vice versa. To make this effect more visible to the subjects, the difference in

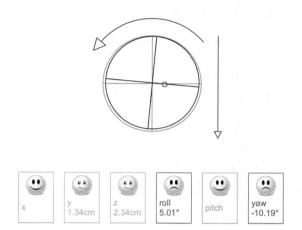

Figure 7.4: The feedback interface for all six degrees of freedom [129].

position is scaled according to Equation (7.4):

$$r = r_0 + 10 \cdot (z_{\text{ref}} - z),$$ (7.4)

with r as the new radius, r_0 as the reference radius, z_{ref} as the reference position and z as the subjects current position. To control the *pitch* and *yaw* orientations, an additional circular marker is added to the interface which indicates the subject's nose. Ideally, this marker should be in the center of the cross air. Figure 7.4 shows the full interface. In addition to the cross airs, arrows are added to help the subject correct in the right directions.

The severity of the deviations in all degrees of freedom are represented by smiley faces, ranging from happy over worried to sad.

7.1.4 Tolerances

To give sensible feedback, tolerances for both position and orientation have to be defined. These tolerances are chosen to be the maximum offset tolerable to the measurement. For the threshold in orientation angles, values are taken from the related work by Denk [126]. These values, 0.8°for roll and pitch, and 0.6°for the

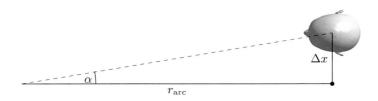

Figure 7.5: Sketch of the angle influence α from a position offset Δx of the subjects head.

yaw angle, can be validated using the Just Noticeable Difference (JND) values for the localization blur discussed in Section 2.4.2. The work of Blauert gives JND values between $0.7\,°$ and $4\,°$. The chosen tolerance values lie on the lower estimation which should result in non-audible differences if the movement is restricted within these tolerances.

The position tolerance values are chosen by Denk as 1 cm for all three room axes. To translate this tolerance to the minimal audible angle again, the influence of a 1 cm shift in one room direction on the angle is considered. Figure 7.5 shows a sketch of the consideration. The influence on the angle is the largest if the movement is perpendicular to the source position. In that case, the offset angle α resulting from the position offset Δx can be calculated as

$$\alpha = \arctan\left(\frac{\Delta x}{r_{\mathrm{arc}}}\right). \tag{7.5}$$

The resulting maximal angle for a 1 cm offset with the radius of the arc $r_{\mathrm{arc}} = 1.2m$ is $\alpha = 0.48°$. This also lies well below the JND values summarized by Blauert.

Within these tolerances, the feedback display shows the happy smiley, indicating that everything is in order. If the deviation lies between the tolerance, and twice the tolerance for orientation or 3,5 times the tolerance for position, the smiley changes to worried. Above these values the smiley changes to sad.

7.1.5 Subject Orientation Results

This section presents the measured orientation results for all four cases. As the observations are very similar for all three orientation angles, only the pitch values are discussed in detail. Roll and Yaw results can be found in Appendix B.1.

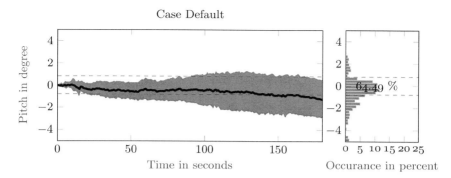

Figure 7.6: Pitch movement over time for all subjects for case *Default*. The left figure shows mean and standard deviation over the subjects. The right figure shows a histogram of all tracking information and all subject measurements.

Figure 7.6 shows the deviations in pitch orientation for all subjects for the default measurement setup. In the left plot the pitch orientation over time is shown as a mean and standard deviation for all subjects. The right plot additionally shows a histogram of the yaw values of all subjects. Both plots additionally show a tolerance limit, discussed in Section 7.1.4. A clear deviation from those tolerances can be observed with the standard deviation between subjects increasing over time. The mean deviation increases to approximately 2°at the end of the measurement. This might be a result from fatigue caused by standing still over extended time periods, even though the overall measurement time of three minutes can be considered relatively short. Similar results can be seen also for the turntable case seen in Figure 7.7. Here, slight post-processing artifacts can be found in the mean value, which increases from 40 s to 60 s but generally shows the same decrease in pitch as seen before. Overall, the subjects are inside the tolerance limits only for 48 % of the time compared to 64 % in the default case, where the subject is not rotated. In the case of sitting subjects, this fatigue effect is not as pronounced. Figure 7.8 shows the expected increase in standard deviation while the mean value remains relatively constant and within the tolerance limits. With this measurement setup, the subjects are within tolerances 76 % of the time. Even greater values are achieved using the feedback system described in Section 7.1.3. In this case, a different movement behavior over time is measured. In the beginning of the measurement, the standard deviation is largest with values of about 3°. This deviation decreases after approximately 30 s and the subjects all remain inside the tolerance levels for the remaining measurement duration.

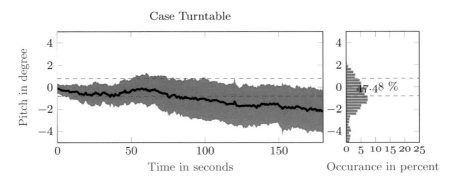

Figure 7.7: Pitch movement over time for all subjects for case *Turntable*. The left figure shows mean and standard deviation over the subjects. The right figure shows a histogram of all tracking information and all subject measurements.

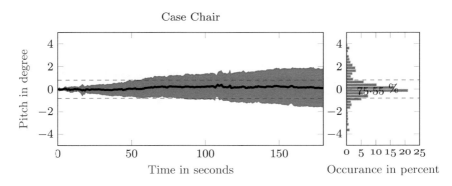

Figure 7.8: Pitch movement over time for all subjects for case *Chair*. The left figure shows mean and standard deviation over the subjects. The right figure shows a histogram of all tracking information and all subject measurements.

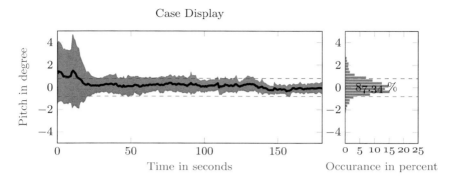

Figure 7.9: Pitch movement over time for all subjects for case Display. The left Figure shows mean and standard deviation over the subjects. The right Figure shows a histogram of all tracking information and all subject measurements.

Overall 87 % of values lie within the tolerance limits. As this case is the only feedback case, two conclusions can be drawn. The first conclusion is favorable to the system. Once subjects got used to the feedback, they could stay inside the shown tolerances. The second conclusion is that the given training time and method are not suitable to fully master the system and reduce deviations. Even after the employed training, the subjects are able to fully control the pitch movement only after an additional 30 s.

Table 7.2 summarizes the percentage of time that subjects are within the tolerances for all three orientation degrees of freedom for all four measurement cases. Note that the value for the pitch deviation in the turntable case is highly affected by the post-processing and does not reflect the subject's performance accurately.

Table 7.2: Time in percent that the subjects are inside the given tolerance levels for all three orientation degrees of freedom for all four measurement cases.
* This value does not reflect the subject's performance accurately.

	Default	Turntable	Chair	Display
Roll	87.65 %	83.60 %	89.57 %	96.18 %
Pitch	64.49 %	47.48 %*	75.55 %	87.34 %
Yaw	94.98 %	60.47 %	82.23 %	88.02 %

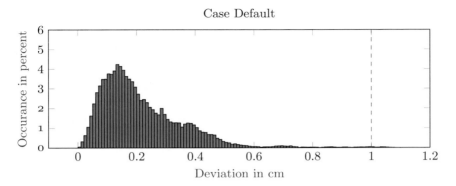

Figure 7.10: Histogram of all subjects combined position offset in the default case. The tolerance value of the display case at 1 cm is indicated as the dashed line.

7.1.6 Subject Position Results

This section shows the analysis of the subjects position mismatch. As the data shows that the subjects did not move beyond the tolerance value of 1 cm very often for any direction or measurement case, the analysis is only presented in a compact fashion. Figure 7.10 shows a histogram of subject position errors for the default case. The errors are calculated as the length of the global offset vector, combining all three dimensions x, y and z. Equation (7.6) shows the calculation of the combined offset value n:

$$n = \sqrt{x^2 + y^2 + z^2}. \tag{7.6}$$

This simplification is done as no real gain can be made from a selective examination. Indeed, the measurement setup does not have an influence on the extent of the subject's mis-positioning. The plots for the three other cases can be found in Appendix B.2. The Figure shows the maximum global position offset at around 0.1 cm. This low position offset can be attributed to the use of the head-rest which limits position movements effectively. The influence of the head-rest has already been evaluated with both rotating and standing subjects [128]. Here, very similar results have been found.

Overall, the influence of the position mismatch on the measured angle can be neglected compared to the angle mismatch itself.

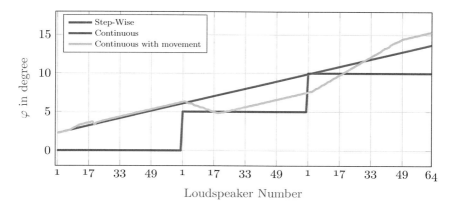

Figure 7.11: Visualization of the influence of subject movement on the relative azimuth angle between source and subject. As a reference, an undisturbed continuous and a traditional step-wise measurement are shown.

7.2 Error from Movement

This section describes the correction method developed to combat subject movement during HRTF measurements. The method is similar to the continuous movement correction and works as an add-on to the previously described principles.

Again, the correct relative angles between the loudspeakers and the microphones are needed. To this end, an optical head tracking system is used. From this system, a relative position can be extracted.

7.2.1 Position Correction

Figure 7.11 shows the influence of subject movement during the measurement, analogously to Figure 6.1. A clear mismatch between the expected (━━) and real (━━) relative angle can be seen. To calculate the correct relative angle both position and orientation of the subject have to be taken into consideration.

Assuming the loudspeaker on a circular path around the coordinate system's origin, a position deviation by the subject can be corrected by subtracting the

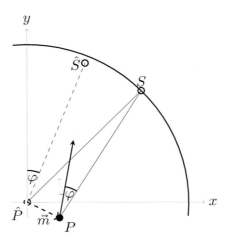

Figure 7.12: Sketch of the position correction. The angle and distance between the source position S needs to be calculated relative to subject P, considering the subject position and orientation. A corrected subject position $\hat{P}$ and corrected sources position $\hat{S}$ are obtained.

offset from both the subject and the loudspeaker:

$$\hat{P} = P - \vec{m}, \tag{7.7}$$

with the new loudspeaker position $\hat{P}$ as a result of the real loudspeaker position P minus the subject's offset from the origin $\vec{m}$.

The real loudspeaker position $\hat{S}$ is not only influenced by the subject's positional offset but also by its orientation. To account for this, the rotation matrix $\mathbf{A}$ derived from the roll, pitch and yaw angles, representing rotation around x, y and z, is calculated and its inverse is multiplied to the position-corrected source position.

$$\hat{S} = \mathbf{A}^{-1} \cdot (S - \vec{m}). \tag{7.8}$$

Figure 7.12 sketches the correction for position and orientation in two dimensions. As the subject moves towards or away from the sound source, the distance changes. These changes in radius are neglected in the following as they can be considered very small compared to the radius itself. Further improvements of the algorithm could include a phase correction according to this positional offset. As a result of the correction, the movement heavily impacts the sampling grid of the measured HRTF. As the movement also occurs during the sweep measurement, a frequency-dependent influence can be expected, similar to the influence of the continuous measurement, described in Chapter 6. For the same reasons a spatial interpolation is needed to correct the frequency-dependent shift. Again, a spherical harmonic decomposition is used to that end.

As the quality of the decomposition and reconstruction is highly dependent on the sampling grid [103], even with the applied regularization, the correction of the subject's movements is critical for the quality of the transformation. The movements can, for example, result in two measurements with the same spatial position.

To account for this, a linear interpolation between all measurement points per elevation slice is applied before the spherical harmonics transformation. This interpolation reduces the condition number of the transformation matrix significantly and thus reduces additional error from the transformation.

7.2.2 Measurement Data

To quantize the influence of azimuth movement on HRTF measurements, five datasets of the HRTF of an artificial head are acquired. The measurements are obtained with continuous system rotation around the head. The head itself is positioned on a turntable and is rotated in azimuth during the measurement procedure to simulate a subject's yaw movement. While one reference measurement is taken without subject movement, the remaining four measurements contain increasing amounts of movement. The artificial head is moved with random speeds between $1\,°/s$ and $8\,°/s$, while the angle is generated from a normal distribution around $0\,°$. To increase the amount of movement between measurements, the standard deviation of this normal distribution is increased. This increases the highest achievable angle while still maintaining an average of $0°$. The chosen standard deviation values are $\sigma = 0.25\,°, 0.5\,°, 1\,°, 2\,°$. The random head rotations are done consecutively, a new rotation is started immediately after the old rotation is finished. In the following the five measurements are addressed by the standard deviation of the normal distribution of the measurement. The bigger this value, the more movement is expected during the measurement.

To validate the movement and the normal distribution of the angles, the tracked data of three measurements is exemplary shown as a histogram in Figure 7.13. The histogram shows the occurrence of larger deviation values if the standard

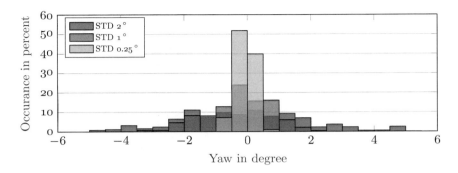

Figure 7.13: Histogram of the occurrence of the deviation in φ from three measurements.

deviation is larger. For the smallest standard deviation, the deviation angle is no larger than $1°$.

7.2.3 Measurement Error

Figure 7.14 shows the spectral difference error according to Equation (4.1). The uncorrected error shows expected results. With increasing frequency, the error is increasing. As the movement is larger compared to the wavelength for higher frequencies, this tendency is expected. Secondly, the error is increasing with more movement. This also expected trend shows the validity of the assumption that more movement is detrimental to the measurement quality.

The overall error is reduced using the correction of the measurement with the acquired tracker positions. While the error exceeds $2\,\text{dB}$ at higher frequencies for the STD $2°$ measurement (- - -), it is reduced by approximately $1\,\text{dB}$ by applying the correction (——).

As a reference for the error a second reference measurement is taken and is compared to a repetition error of the measurement setup (——). The error between the measurements lies below $1\,\text{dB}$ over the whole frequency range.

Using the correction, the error of the STD 1 measurement (- - -) is reduced to approximately this repetition level, indicating a sufficient error reduction (——).

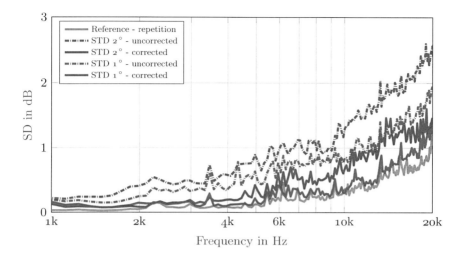

Figure 7.14: Spectral difference values for movement-disturbed measurements. Measurements with STD 2° and STD 1° for both uncorrected and corrected are shown. As a reference, the measurement setup repetition error is shown.

The Figure does not show errors for other measurement types with less disturbance as these errors do not significantly differ from the repetition error.

The influence of the correction on the data can also be seen in Figure 7.15. The figure shows both magnitude and phase deviations from the measurement with STD = 2° movement relative to the reference for the horizontal plane at 10000 Hz. The data for the left ear is shown only. A clear reduction in deviation can be seen as a result of the applied correction. Additionally, the spatial distribution of the errors over the sphere is exemplary shown. For the hemisphere facing the source $(0° \leq \varphi < 180°)$ the errors are smaller compared to the other, contra-lateral, hemisphere. Similar behavior is already discussed in Section 6.5.2 and can be explained by the distribution of spatial resonances over the HRTF sphere. At the contra-lateral side, a higher density of dips can be found. These dips result in larger errors, if not correctly sampled.

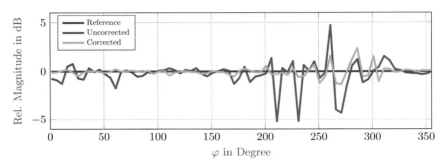

(a) Amplitude deviation relative to reference.

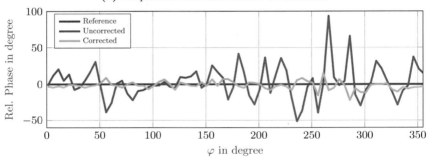

(b) Phase deviation relative to reference.

Figure 7.15: Influence of the movement (STD $= 2°$) on single measurements. The figures show both amplitude and phase deviations from the reference for the horizontal plane of the left ear measurements at 10000 Hz.

7.3 Summary

This chapter investigated subject movement during HRTF measurements. For most publications, the movement is considered detrimental to the quality of the resulting measurements and are therefore avoided.

Two investigations into the movement were presented. The first investigation looks into the **amount** of movement subjects do depending on the measurement setup. Four different setups were investigated which study standing or sitting position, subject rotation and a live feedback system. While the position deviation were similar and neglect-able for all setup types, the orientation of the subject was more controlled in a sitting position or in standing position with the use of the feedback display. With the use of the feedback system, all deviations in the three orientation degrees of freedom are limited to a maximum of $1\,°$ after a training time.

In the second part, a preliminary investigation into the **influence** of movement onto measurement quality was done. For this investigation, an artificial head was measured using a measurement setup that was continuously rotating around the head. During the measurement, the head was rotated around its axis to simulate a subject's *yaw* movement. The amount of movement was controlled and increased over four consecutive measurements. A correction for the movement has been proposed that works analogously to the correction for continuous rotation proposed in Chapter 6. The correction reduced the error successfully. Even for relatively large movements in the STD $1\,°$ case, in which the deviations are up to $2.5\,°$, the error was reduced to the same level as a repetition error of the measurement system.

This improvement, in combination with a feedback system during the measurement, can be used to eliminate the influence of the yaw movement. As a future study, both roll and pitch movements should be examined separately and in combination to determine more realistic boundaries of the movement correction.

8

Conclusion and Outlook

This thesis presented an objective and subjective evaluation of a measurement system for the fast acquisition of spatially dense, individual Head-Related Transfer Function (HRTF).

The system is a progression of the system previously developed at the Institute of Technical Acoustics (ITA) by Masiero [96]. After the previous system suffered hardware failure, a new construction was preferred to remedy known drawbacks in the design of the old system. The drawbacks included the directivity of the used loudspeaker, reflections between loudspeaker housings, the inability to rotate the setup itself and the difficulty to enter the measurement system prohibiting participants with limited mobility to be measured. An additional goal of the new system was to be as small as possible to reduce the influence of the system on the sound-field.

In the first part of the thesis an objective evaluation of the measurement system was presented. The evaluation was performed by conducting measurements of objects with different spatial complexity. To compare measurement data with both a frequency and a spatial dependency, different error measures have been defined to simplify visualization of the error. The introduced error measures average over the spatial sphere and return a frequency-dependent error value. As a first test, the sound-field inside the arc was measured using multiple microphones on a linear array. From this data, an approximation of the directivity of one loudspeaker was extracted. As expected, more disturbances in the sound field are visible, compared to a directivity measurement of a traditional HRTF measurement system. These disturbances start at approximately 7 kHz but the sound pressure variation mostly lies within 3 dB compared the frontal direction. These measurements indicate a slightly limited measurement quality when using the fast HRTF measurement system. As a second measurement, a solid sphere with a microphone on the surface was used. This device was chosen as a defined analytical solution is available to compare against. The diameter of the sphere was approximately the same as a typical human head. Using this object, the influence of the measurement setup on a head can be approximated. The comparison showed overall good agreements with a reference measurement. Both measurements showed errors of the same magnitude when comparing to an analytical solution. In agreement with the directivity measurement, the error increases starting at approximately 7 kHz. As a third evaluation an artificial head was measured in both the fast measurement system and a reference system. Additionally, a BEM simulation of the head was used to quantify the error. Again, the comparison between the measurements show a very good agreement until 7 kHz. Starting at this point, the spectral differences increass slightly and reach approximately 3 dB at maximum. At frequencies over 15 kHz the error is again increasing slightly. The comparison against the BEM simulation did not show good agreement. While there were some mismatch between measured and simulated data visible, the used error measures were shown to not ideally represent the differences for comparisons against simulated data. An additional comparison with data from the old measurement system showed errors in the same magnitude, although slightly elevated compared to the new system.

To examine whether these differences are detrimental to the quality of binaural synthesis, several listening experiments have been conducted and were used as a subjective evaluation of the system's quality. The first two experiments, performed using measurements from the previous measurement system, studied the quality of the acquired HRTFs with regard to localization accuracy. No significant differences between localization ability using HRTFs from a fast, and using HRTFs from a traditional measurement system could be detected indicating no detectable loss in quality from the fast measurement system. Moreover, individual HRTFs, compared significantly better than non-individual HRTFs and similar accuracy levels as in previous literature were achieved suggesting the quality of the HRTF to be sufficient. In a third experiment, using the new fast measurement system, similar results were achieved. In this experiment, additional to the previously used static reproduction, a dynamic sound source reproduction was tested. Using this reproduction method, significantly decreased localization and front-back confusion error rates were achieved which was again predicted by previous research indicated.

The findings from both objective and subjective evaluation indicate that the system works satisfyingly well and the quality of the acquired HRTFs is of comparable quality to other measurement methods, despite errors introduced from the measurement system as seen during the objective evaluation.

To further improve the measurement time, an additional measurement mode was introduced and evaluated. Using this measurement method, the pauses between measurements in which the subject is usually re-positioned to a new measurement angle are removed. The measurement signal plays continuously and a continuous slow rotation is employed. A method to correct this motion from the measurements was introduced. Furthermore, an objective and subjective evaluation of the method was presented. These evaluations showed that the introduced error from the continuous rotation decreases with the rotation speed. A subjective evaluation with the goal of finding the fastest measurement that does not introduce audible differences was performed. The overall audibility of the stimuli was low. Only with the fastest rotational speeds, the subjects could detect audibility changes consistently. Based on these findings, the rotational speed could be chosen to shorten the measurement from approximately seven minutes to approximately three minutes.

Lastly, an evaluation of the amount of subject movement during different measurement setup situations and their influence on the measurement quality was presented. Four different situations were evaluated, comparing standing and sitting subjects, differences between rotation of the subject itself compared to a measurement where the measurement setup is rotated. Additionally, a live feedback system was employed during one test that lets the subjects control their own position and orientation. Using this feedback system, the subjects were able to maintain their position in-between defined limits within $1°$ after a short training time. Without this feedback system, the subjects best performed while seated. The most deviation from the desired position occurred during the subject rotation.

Using the previously introduced correction methods needed for continuous measurement rotation, an evaluation on the magnitude of movement that could still be corrected was presented. This evaluation used several measurements of an artificial head that was rotated around one axis during a simultaneous measurement. As expected, the introduced error increased with the introduced movement. It was shown that if the movement has been limited to $1°$ in yaw, the errors after the correction were not larger than a repetition error of one measurement. Combined with a measurement using visual feedback, the presented methods show promise to remove all influence of the movement from the measurement, however, an evaluation of movement in the other two orientation directions and the position mismatch have to be studied in more detail.

Outlook

To further improve the methods presented in this thesis, several approaches are possible. The presented error measures, while giving sensible values that correlate somewhat to perception, are basic. Multiple improvements to better depict the HRTF quality are thinkable. It was shown that for comparisons to simulations, or if the room temperature changed between measurements, the error measure fails as slight frequency deviations in resonance frequencies can cause large errors. Here, a correction for resonances that have a non-noticeable frequency shift might improve the error measure. An overall evaluation regarding localization accuracy and audibility of the errors is also necessary to gain insight into the JND of this error measure. Furthermore, the subjective HRTF quality can be evaluated further using approaches suggested by Nicol et al. [130]. A different evaluation could be the comparison of individual measurements against BEM simulations or fast prototyping of the head acquired by 3D imaging methods [131, 132].

The evaluation of influence of the subject's movements are only shown for yaw movements. While it is expected that the influence is of comparable size for roll and pitch, a separate and combined evaluation is of further interest. To this end, an artificial head with controllable head over torso movements in all three degrees of freedom needs to be constructed.

Appendices

A

Objective Evaluation

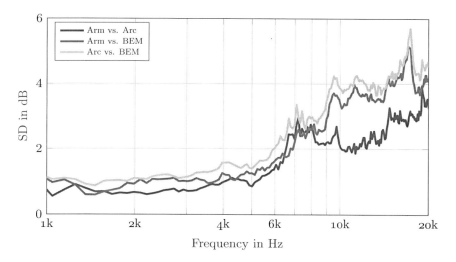

Figure A.1: SD error of the comparison between the traditional measurement system, the fast measurement system and a BEM simulation of the right ear of an artificial head.

B

Subject Movement

B.1 Orientation

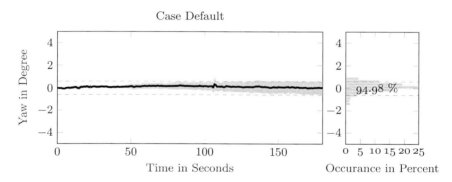

Case Default

Figure B.1: Yaw movement over time for all subjects for case Default. The left Figure shows mean and standard deviation over the subjects. The right Figure shows a histogram of all tracking information and all subject measurements.

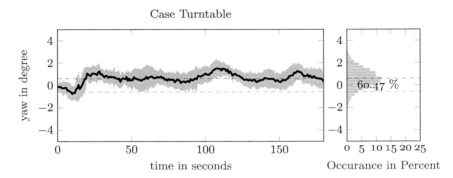

Figure B.2: Yaw movement over time for all subjects for case Turntable. The left Figure shows mean and standard deviation over the subjects. The right Figure shows a histogram of all tracking information and all subject measurements.

1 2

---

[1] Note, that the yaw movement in the turntable case is prone to postprocessing artifacts.
[2] Furthermore, there is a funkte.

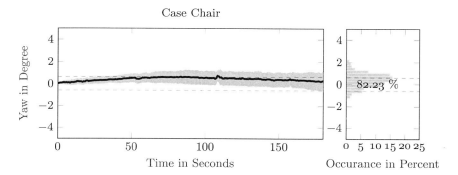

Figure B.3: Yaw movement over time for all subjects for case Chair. The left Figure shows mean and standard deviation over the subjects. The right Figure shows a histogram of all tracking information and all subject measurements.

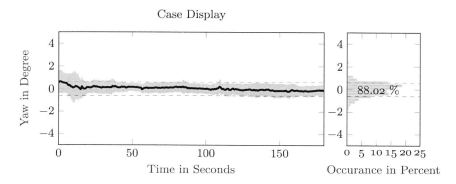

Figure B.4: Yaw movement over time for all subjects for case Display. The left Figure shows mean and standard deviation over the subjects. The right Figure shows a histogram of all tracking information and all subject measurements.

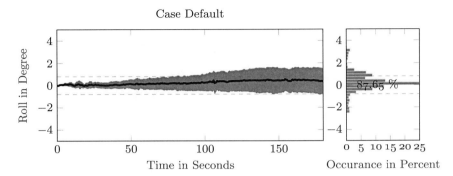

Figure B.5: Roll movement over time for all subjects for case Default. The left Figure shows mean and standard deviation over the subjects. The right Figure shows a histogram of all tracking information and all subject measurements.

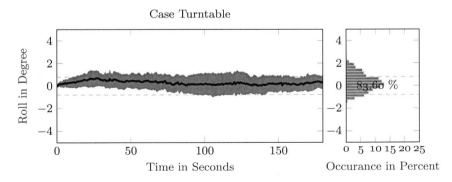

Figure B.6: Roll movement over time for all subjects for case Turntable. The left Figure shows mean and standard deviation over the subjects. The right Figure shows a histogram of all tracking information and all subject measurements.

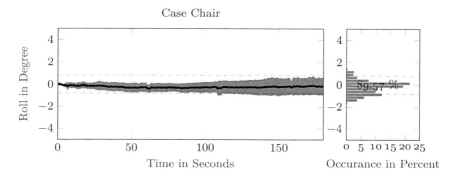

Figure B.7: Roll movement over time for all subjects for case Chair. The left Figure shows mean and standard deviation over the subjects. The right Figure shows a histogram of all tracking information and all subject measurements.

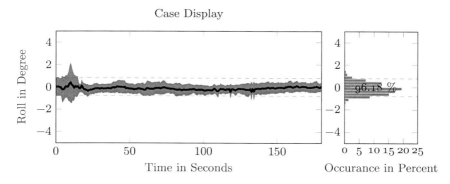

Figure B.8: Roll movement over time for all subjects for case Display. The left Figure shows mean and standard deviation over the subjects. The right Figure shows a histogram of all tracking information and all subject measurements.

B.2 Position

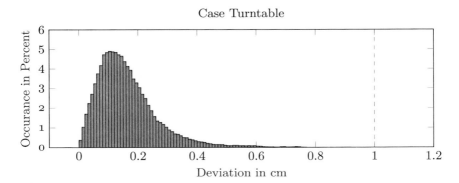

Figure B.9: Histogram of all subjects combined position offset in the turntable case. The tolerance value of the display case at 1 cm is indicated as the dashed line.

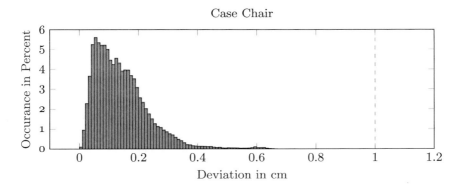

Figure B.10: Histogram of all subjects combined position offset in the chair case. The tolerance value of the display case at 1 cm is indicated as the dashed line.

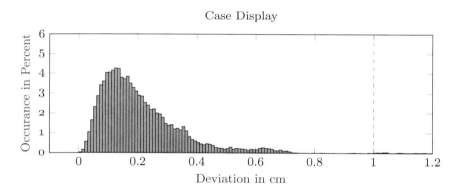

Figure B.11: Histogram of all subjects combined position offset in the display case. The tolerance value of the display case at 1 cm is indicated as the dashed line.

List of Acronyms

BEM Boundary Element Method

BRIR Binaural Room Impulse Response

DSHT Discrete Spherical Harmonic Transformation

DTF Directional Transfer Function

GELP God's Eye View Localization Point

HMD Head Mounted Display

HpTF Headphone Transfer Function

HRTF Head-Related Transfer Function

ITA Institute of Technical Acoustics

ILD Interaural Level Difference

ITD Interaural Time Difference

JND Just Noticeable Difference

LMS Least Mean Squares

LTI Linear Time-Invariant

SD Spectral Differences

SHT Spherical Harmonic Transformation

SNR Signal to Noise Ratio

3AFC Three-Alternative Forced-Choice

List of Figures

List of Tables

Bibliography

[1] M. Morimoto and Y. Ando. "On the simulation of sound localization." In: *J. Acoust. Soc. Japan* 1.3 (1980), pp. 167–174. ISSN: 0388-2861. DOI: 10.1250/ast.1.167 (cit. on pp. 1, 15, 16).

[2] J. Fels, P. Buthmann, and M. Vorländer. "Head-related transfer functions of children." In: *Acta Acust. united with Acust.* 90.5 (2004), pp. 918–927. ISSN: 16101928 (cit. on p. 1).

[3] H. Møller et al. "Binaural technique: Do we need individual recordings?" In: *J. Audio Eng. Soc* 44.6 (1996), pp. 451–469 (cit. on pp. 1, 18).

[4] D. R. Begault and E. M. Wenzel. "Headphone localization of speech." In: *Hum. Factors* 35.2 (1993), pp. 361–376. ISSN: 00187208. DOI: 10.1177/001872089303500210 (cit. on pp. 1, 18, 64).

[5] J. Oberem et al. "Intentional Switching in Auditory Selective Attention: Exploring Different Binaural Reproduction Methods in an Anechoic Chamber." In: *Acta Acust. united with Acust.* 100.6 (2014), pp. 1139–1148. ISSN: 16101928. DOI: 10.3813/AAA.918793 (cit. on p. 1).

[6] H. Wallach. "The Role of Head Movements and Vestibular and Visual Cues in Sound Localization." In: *English* 27.4 (1940), pp. 339–368. ISSN: 0022-1015. DOI: 10.1037/h0054629 (cit. on pp. 2, 20).

[7] W. R. Thurlow, J. W. Mangels, and P. S. Runge. "Head Movements During Sound Localization." In: *J. Acoust. Soc. Am.* 42.2 (1967), pp. 489–493. ISSN: 0001-4966. DOI: 10.1121/1.1910605 (cit. on pp. 2, 20).

[8] I. Pollack and M. Rose. "Effect of head movement on the localization of sounds in the equatorial plane." In: *Percept. Psychophys.* 2.12 (1967), pp. 591–596 (cit. on pp. 2, 20).

[9] L. Rayleigh. "On our perception of sound direction." In: *Philos. Mag. Ser. 6* 13.74 (1907), pp. 214–232. ISSN: 1941-5982. DOI: 10.1080/14786440709463595 (cit. on pp. 3, 4, 17).

[10] M. Vorländer. *Auralization: Fundamentals of acoustics, modelling, simulation, algorithms and acoustic virtual reality*. Springer-Verlag, 2007. ISBN: 9783540488293 (cit. on pp. 4, 9).

[11] D. M. Leakey, B. M. Sayers, and C. Cherry. "Binaural Fusion of Low- and High-Frequency Sounds." In: *J. Acoust. Soc. Am.* 30.3 (1958), pp. 222–222. ISSN: 0001-4966. DOI: 10.1121/1.1909549 (cit. on p. 6).

[12] G. B. Henning. "Detectability of interaural delay in high-frequency complex waveforms." In: *J. Acoust. Soc. Am.* 55.1 (1974), pp. 84–90. ISSN: 0001-4966. DOI: 10.1121/1.1928135 (cit. on p. 6).

[13] D. McFadden and E. G. Pasanen. "Lateralization at high frequencies based on interaural time differences." In: *J. Acoust. Soc. Am.* 59.3 (1976), pp. 634–639. ISSN: 0001-4966. DOI: 10.1121/1.380913 (cit. on p. 6).

[14] E. M. von Hornbostel and M. Wertheimer. "Über die Wahrnehmung der Schallrichtung." In: *Sitzungsberichte der Preuss. Akad. der Wissenschaften* 388 (1920), p. 396 (cit. on p. 6).

[15] J. Blauert. *Spatial hearing: The psychophysics of human sound localization*. 2nd. MIT Press Cambridge, 1997 (cit. on pp. 6, 7, 14, 17, 20).

[16] N. J. Wade and D. Deutsch. "Binaural Hearing—Before and After the Stethophone." In: *Acoust. Today* 4.3 (2008), p. 16. ISSN: 1557-0215. DOI: 10.1121/1.2994724 (cit. on p. 8).

[17] S. Paul. "Binaural recording technology: A historical review and possible future developments." In: *Acta Acust. united with Acust.* 95.5 (2009), pp. 767–788. ISSN: 16101928. DOI: 10.3813/AAA.918208 (cit. on p. 8).

[18] H. Møller. "Fundamentals of binaural technology." In: *Appl. Acoust.* 36.3-4 (1992), pp. 171–218. ISSN: 0003682X. DOI: 10.1016/0003-682X(92)90046-U (cit. on p. 8).

[19] F. Wefers. "Partitioned convolution algorithms for real-time auralization." PhD thesis. RWTH Aachen University, 2015, IV, 258 S. : Ill., graph. Darst. ISBN: 978-3-8325-3943-6 (cit. on p. 8).

[20] B. Masiero and J. Fels. "Perceptually robust headphone equalization for binaural reproduction." In: *Audio Eng. Soc. Conv. 130* (2011), pp. 1–7 (cit. on pp. 9, 63).

[21] J. Oberem, B. Masiero, and J. Fels. "Experiments on authenticity and plausibility of binaural reproduction via headphones employing different recording methods." In: *Appl. Acoust.* 114 (2016), pp. 71–78. ISSN: 1872910X. DOI: 10.1016/j.apacoust.2016.07.009 (cit. on pp. 10, 11, 15, 39).

[22] M. D. Burkhard and R. M. Sachs. "Anthropometric manikin for acoustic research." In: *J. Acoust. Soc. Am.* 58.1 (1975), pp. 214–222. ISSN: 0001-4966. DOI: 10.1121/1.380648 (cit. on p. 11).

[23] A. Schmitz. "Ein neues digitales Kunstkopfmeßsystem." In: *Acta Acust. united with Acust.* 81 (1995), pp. 416–420 (cit. on p. 11).

[24] H. Møller et al. "Head-related transfer functions of human subjects." In: *J. Audio Eng. Soc* 43.5 (1995), pp. 300–321 (cit. on pp. 11, 15, 93, 98).

[25] E. M. Wenzel et al. "Localization using nonindividualized head-related transfer functions." In: *J. Acoust. Soc. Am.* 94.1 (July 1993), pp. 111–123. ISSN: 0001-4966. DOI: 10.1121/1.407089 (cit. on pp. 11, 18, 71).

[26] J. C. Middlebrooks. "Virtual localization improved by scaling nonindividualized external-ear transfer functions in frequency." In: *J. Acoust. Soc. Am.* 106.3 Pt 1 (1999), pp. 1493–1510. ISSN: 00014966. DOI: 10.1121/1.427147 (cit. on pp. 11, 18, 61, 66).

[27] A. W. Bronkhorst. "Localization of real and virtual sound sources." In: *J. Acoust. Soc. Am.* 98.November 1995 (1995), pp. 2542–2553 (cit. on pp. 11, 15, 18, 93, 98).

[28] V. Larcher, J.-M. Jot, and G. Vandernoot. "Equalization Methods in Binaural Technology." In: *105th AES Conv.* Vol. 4858. 1998, Convention Paper 4858 (cit. on p. 12).

[29] J. C. Middlebrooks and D. M. Green. "Directional dependence of interaural envelope delays." In: *J. Acoust. Soc. Am.* 87.5 (1990), pp. 2149–2162. ISSN: 0001-4966. DOI: 10.1121/1.399183 (cit. on p. 12).

[30] D. J. Kistler and F. L. Wightman. "A model of head-related transfer functions based on principal components analysis and minimum-phase reconstruction." In: *J. Acoust. Soc. Am.* 91.3 (1992), pp. 1637–1647. ISSN: 0001-4966. DOI: 10.1121/1.402444 (cit. on p. 12).

[31] G. W. Stewart. "The acoustic shadow of a rigid sphere, with certain applications in architectural acoustics and audition." In: *Phys. Rev. (Series I)* 33.6 (1911), pp. 467–479. ISSN: 0031899X. DOI: 10.1103/PhysRevSeriesI.33.467 (cit. on p. 12).

[32] G. W. Stewart. "Phase relations in the acoustic shadow of A rigid sphere; Phase difference at the ears." In: *Phys. Rev.* 4.3 (1914), pp. 252–258. ISSN: 0031899X. DOI: 10.1103/PhysRev.4.252 (cit. on p. 12).

[33] D. S. Brungart, W. M. Rabinowitz, and N. I. Durlach. "Auditory localization of a nearby point source." In: *J. Acoust. Soc. Am.* 100.4 (1996), p. 2593. DOI: 10.1121/1.417577 (cit. on p. 12).

[34] R. O. Duda and W. L. Martens. "Range dependence of the response of a spherical head model." In: *J. Acoust. Soc. Am. J. Acoust. Soc. Am. J. Acoust. Soc. Am.* 104.111 (1998), pp. 1465–547. DOI: 10.1121/1.407089 (cit. on p. 12).

[35] D. S. Brungart and W. M. Rabinowitz. "Auditory localization of nearby sources. Head-related transfer functions." In: *J. Acoust. Soc. Am.* 106.3 Pt 1 (1999), pp. 1465–1479. ISSN: 00014966. DOI: 10.1121/1.427180 (cit. on p. 12).

[36] A. Farina. "Simultaneous measurement of impulse response and distortion with a swept-sine technique." In: *Proc. AES 108th conv, Paris, Fr.* I (2000), pp. 1–15. DOI: 10.1109/ASPAA.1999.810884 (cit. on p. 13).

[37] S. Müller and P. Massarani. "Transfer-function measurement with sweeps." In: *J. Audio Eng. Soc.* 49.6 (2001), pp. 443–471 (cit. on p. 13).

[38] C. L. Searle et al. "Binaural pinna disparity: another auditory localization cue." In: *J. Acoust. Soc. Am.* 57.2 (1975), pp. 448–455. ISSN: 00014966. DOI: 10.1121/1.380442 (cit. on pp. 15, 93).

[39] S. Mehrgardt and V. Mellert. "Transformation characteristics of the external human ear." In: *J. Acoust. Soc. Am.* 61.6 (1977), pp. 1567–1576. ISSN: 0001-4966. DOI: 10.1121/1.381470 (cit. on p. 15).

[40] E. M. Wenzel, F. L. Wightman, and S. H. Foster. "A Virtual Display System for Conveying Three-Dimensional Acoustic Information." In: *Proc. Hum. Factors Ergon. Soc. Annu. Meet.* Vol. 32(2). 1988, pp. 86–90. DOI: 10.1177/154193128803200218 (cit. on p. 15).

[41] F. L. Wightman and D. J. Kistler. "Headphone simulation of free-field listening. I: Stimulus synthesis." In: *J. Acoust. Soc. Am.* 85.2 (Feb. 1989), pp. 858–867. ISSN: 0001-4966 (cit. on pp. 15, 16, 93).

[42] J. C. Middlebrooks. "Narrow-band sound localization related to external ear acoustics." In: *J. Acoust. Soc. Am.* 92.5 (1992), pp. 2607–2624. ISSN: 0001-4966. DOI: 10.1121/1.404400 (cit. on p. 15).

[43] E. H. A. Langendijk and A. W. Bronkhorst. "Fidelity of three-dimensional-sound reproduction using a virtual auditory display." In: *J. Acoust. Soc. Am.* 107.1 (Dec. 1999), p. 528. ISSN: 0001-4966. DOI: 10.1121/1.428321 (cit. on pp. 15, 21).

[44] E. H. Langendijk and A. W. Bronkhorst. "Contribution of spectral cues to human sound localization." In: *J. Acoust. Soc. Am.* 112.4 (2002), pp. 1583–1596. ISSN: 0001-4966. DOI: 10.1121/1.424945 (cit. on pp. 15, 16).

[45] D. N. Zotkin, R. Duraiswami, and E. Grassi. "Fast head-related transfer function measurement via reciprocity." In: *J. Acoust. Soc. Am.* 120.4 (2006), pp. 2202–2215. ISSN: 00014966. DOI: 10.1121/1.2207578 (cit. on pp. 16, 22).

[46] P. Majdak, P. Balazs, and B. Laback. "Multiple exponential sweep method for fast measurement of head-related transfer functions." In: *J. Audio Eng. Soc* 55.7/8 (2007), pp. 623–637 (cit. on pp. 16, 35).

[47] P. Dietrich, B. Masiero, and M. Vorländer. "On the optimization of the multiple exponential sweep method." In: *J. Audio Eng. Soc* 61.3 (2013), pp. 113–124 (cit. on pp. 16, 35, 79).

[48] B. Masiero, M. Pollow, and J. Fels. "Design of a Fast Broadband Individual Head-Related Transfer Function Measurement System." In: *Proc.. Forum Acust.* Vol. 97. c. Aalborg, Denmark: ACTA ACUSTICA united with ACUSTICA, Hirzel, 2011, pp. 2197–2202. ISBN: 9788469415207 (cit. on p. 16).

[49] G. Enzner. "Analysis and optimal control of LMS-type adaptive filtering for continuous-azimuth acquisition of head related impulse responses." In: *ICASSP, IEEE Int. Conf. Acoust. Speech Signal Process. - Proc.* Vol. 2. 2008, pp. 393–396. ISBN: 1424414849. DOI: 10.1109/ICASSP.2008. 4517629 (cit. on p. 17).

[50] K. Fukudome et al. "The fast measurement of head related impulse responses for all azimuthal directions using the continuous measurement method with a servo-swiveled chair." In: *Appl. Acoust.* 68.8 (2007), pp. 864–884. ISSN: 0003682X. DOI: 10.1016/j.apacoust.2006.09.009 (cit. on p. 17).

[51] V. Pulkki, M.-V. Laitinen, and V. Sivonen. "HRTF measurements with a continuously moving loudspeaker and swept sines." In: *Audio Eng. Soc. Conv. 128.* Audio Engineering Society. 2010, pp. 1–9. ISBN: 9781617387739 (cit. on p. 17).

[52] P. Dietrich et al. "Time Efficient Measurement Method for Individual HRTFs." In: *Fortschritte der Akust. - DAGA 2012.* 2012, pp. 333–334 (cit. on p. 17).

[53] C. Antweiler and G. Enzner. "Perfect sequence lms for rapid acquisition of continuous-azimuth head related impulse responses." In: *Appl. Signal Process. to Audio Acoust. 2009. WASPAA'09. IEEE Work.* Vol. 16. 1. IEEE. 2009, pp. 281–284. ISBN: 0780378504. DOI: 10.1109/LSP.2009. 2015034 (cit. on p. 17).

[54]   M. Fallahi, F. Brinkmann, and S. Weinzierl. "Simulation and analysis of measurement techniques for the fast acquisition of head-related transfer functions." In: 4 (2015), pp. 1107–1110 (cit. on p. 17).

[55]   M. Rothbucher et al. "Comparison of head-related impulse response measurement approaches." In: *J. Acoust. Soc. Am.* 134.2 (Aug. 2013), EL223–EL229. ISSN: 0001-4966. DOI: 10.1121/1.4813592 (cit. on p. 17).

[56]   G. Enzner. "3D-continuous-azimuth acquisition of head-related impulse responses using multi-channel adaptive filtering." In: *IEEE Work. Appl. Signal Process. to Audio Acoust.* 2009, pp. 325–328. ISBN: 9781424436798. DOI: 10.1109/ASPAA.2009.5346532 (cit. on p. 17).

[57]   A. Fuß et al. "Ein vollsphärisches Multikanalmesssystem zur schnellen Erfassung räumlich hochaufgelöster, individueller kopfbezogener Ubertragungsfunktionen." In: *Fortschritte der Akust.* 1 (2015), pp. 666–669 (cit. on p. 17).

[58]   J. R. Angell and W. Fite. "Monaural localization of sound." In: *Psychol. Rev.* 13.333 (1901), pp. 775–777. ISSN: 00368075 (cit. on p. 17).

[59]   S. R. Oldfield and S. P. a. Parker. "Acuity of Sound Localization: A Topography of Auditory Space. I. Normal Hearing Conditions." In: *Perception* 13.5 (1984), pp. 581–600. ISSN: 03010066. DOI: 10.1068/P130581 (cit. on p. 18).

[60]   J. C. Makous and J. C. Middlebrooks. "Two-dimensional sound localization by human listeners." In: *J. Acoust. Soc. Am.* 87.5 (May 1990), pp. 2188–200. ISSN: 0001-4966 (cit. on pp. 18, 19, 62).

[61]   F. L. Wightman and D. J. Kistler. "Headphone simulation of free-field listening. II: Psychophysical validation." In: *J. Acoust. Soc. Am.* 85.2 (1989), pp. 868–878. ISSN: 0001-4966. DOI: 2926001 (cit. on pp. 18, 66).

[62]   M. J. Evans. "Obtaining Accurate Responses in Directional Listening Tests." In: *Audio Eng. Soc. Conv.* 104 730 (1998) (cit. on p. 19).

[63]   R. Mason et al. "Verbal and nonverbal elicitation techniques in the subjective assessment of spatial sound reproduction." In: *J. Audio Eng. Soc.* 49.5 (2001), pp. 366–384. ISSN: 00047554 (cit. on p. 19).

[64]   M. J. Evans, J. A. S. Angus, and A. I. Tew. "Analyzing head-related transfer function measurements using surface spherical harmonics." In: *J. Acoust. Soc. Am.* 104.4 (1998), pp. 2400–2411. ISSN: 0001-4966. DOI: 10.1121/1.423749 (cit. on p. 19).

[65] P. Majdak and B. Laback. "The accuracy of localizing virtual sound sources: Effects of pointing method and visual environment." In: *Audio Eng. Soc. Conv. 124* (2008) (cit. on p. 19).

[66] T. Djelani et al. "An Interactive Virtual-Environment Generator for Psychoacoustic Research II: Collection of Head-Related Impulse Responses and Evaluation of Auditory Localization." In: *Acta Acust. United with Acust.* 86 (2000), pp. 1046–1053. ISSN: 00017884 (cit. on p. 19).

[67] M. a. Frens, A. J. V. Opstal, and R. F. V. D. Willigen. "Spatial and temporal factors determine auditory-visual interactions in human saccadic eye movements." In: *Percept. Psychophys.* 57.6 (1995), pp. 802–816. ISSN: 0031-5117. DOI: 10.3758/BF03206796 (cit. on p. 19).

[68] J. Lewald, G. J. Dörrscheidt, and W. H. Ehrenstein. "Sound localization with eccentric head position." In: *Behav. Brain Res.* 108.2 (2000), pp. 105–125. ISSN: 01664328. DOI: 10.1016/S0166-4328(99)00141-2 (cit. on p. 19).

[69] K. I. McAnally and R. L. Martin. "Sound localization with head movement: Implications for 3-d audio displays." In: *Front. Neurosci.* 8.8 JUL (2014), p. 210. ISSN: 1662453X. DOI: 10.3389/fnins.2014.00210 (cit. on p. 19).

[70] T. Djelani et al. "An Interactive Virtual-Environment Generator for Psychoacoustic Research II: Collection of Head-Related Impulse Responses and Evaluation of Auditory Localization." In: (2000), pp. 1046–1053 (cit. on p. 19).

[71] P. Majdak, M. J. Goupell, and B. Laback. "3-D localization of virtual sound sources: Effects of visual environment, pointing method, and training." In: *Attention, Perception, Psychophys.* 72.2 (2010), pp. 454–469. DOI: 10.3758/APP (cit. on p. 19).

[72] H. Bahu et al. "Comparison of different egocentric pointing methods for 3D sound localization experiments." In: *Acta Acust. united with Acust.* 102.1 (2016), pp. 107–118. ISSN: 16101928. DOI: 10.3813/AAA.918928 (cit. on pp. 19, 70).

[73] R. H. Gilkey et al. "A pointing technique for rapidly collecting localization responses in auditory research." In: *Behav. Res. Methods, Instruments, Comput.* 27.1 (1995), pp. 1–11. ISSN: 07433808. DOI: 10.3758/BF03203614 (cit. on p. 19).

[74] J.-M. Pernaux, M. Emerit, and R. Nicol. "Perceptual Evaluation of Binaural Sound Synthesis: The Problem Of Reporting Localization Judgements." In: (2003) (cit. on p. 19).

[75] J. Braasch and K. Hartung. "Localization in the Presence of a Disctracter and Reverberation in the Frontal Horizontal Plane. I. Psychoacoustical Data." In: *Acta Acust. united with Acust.* 88 (2002), pp. 942–955 (cit. on p. 19).

[76] J. Otten. "Factors influencing acoustical localization." Dissertation. 2001 (cit. on p. 19).

[77] D. Hammershøi and J. Sandvad. "Binaural Auralization. Simulating Free Field Conditions by Headphones." In: *Audio Eng. Soc. Conv. 96.* 1996 (cit. on p. 19).

[78] D. R. Begault, E. M. Wenzel, and M. R. Anderson. "Direct comparison of the impact of head tracking, reverberation, and individualized head-related transfer functions on the spatial perception of a virtual speech source." In: *J. Audio Eng. Soc.* 49.10 (Oct. 2001), pp. 904–916. ISSN: 0004-7554 (cit. on pp. 19, 20, 64).

[79] J.-G. Richter and J. Fels. "Evaluation of localization accuracy of static sources using HRTFs from a fast measurement system." In: *Acta Acust. united with Acust.* 102.4 (2016), pp. 763–771. ISSN: 16101928. DOI: 10.3813/AAA.918992 (cit. on pp. 19, 58, 60, 64, 66, 67).

[80] D. Setzer. "Investigations on head movements in localization experiments with different dynamic binaural reproduction methods." Masterthesis. RWTH Aachen University, 2017 (cit. on p. 19).

[81] P. T. Young. "The Role of Head Movements in Auditory Localization." In: XIV.2 (1931), pp. 579–585 (cit. on p. 20).

[82] F. L. Wightman and D. J. Kistler. "Resolution of front–back ambiguity in spatial hearing by listener and source movement." In: *J. Acoust. Soc. Am.* 105.5 (Apr. 1999), pp. 2841–2853. ISSN: 0001-4966. DOI: 10.1121/1.426899 (cit. on p. 20).

[83] A. W. Mills. "On the Minimum Audible Angle." In: *J. Acoust. Soc. Am.* 30.4 (1958), pp. 237–246. ISSN: 0001-4966. DOI: 10.1121/1.1909553 (cit. on p. 20).

[84] P. Minnaar, J. Plogsties, and F. Christensen. "Directional resolution of head-related transfer functions required in binaural synthesis." In: *J. Audio Eng. Soc* (2005), pp. 919–929 (cit. on pp. 21, 90).

[85] A. V. Oppenheim. *Discrete-time signal processing.* Pearson Education India, 1999 (cit. on p. 22).

[86]   J.-G. Richter et al. "Spherical Harmonics Based HRTF Datasets: Implementation and Evaluation for Real-Time Auralization." In: *Acta Acust. united with Acust.* 100.4 (2014), pp. 667–675. DOI: doi:10.3813/AAA.918746 (cit. on pp. 22, 26).

[87]   E. G. Williams. *Fourier Acoustics - Sound Radiation and Nearfield Acoustical Holography*. Academic Press, 1999. ISBN: 0127539603 (cit. on pp. 22, 23).

[88]   F. Zotter. "Analysis and Synthesis of Sound-Radiation with Spherical Arrays." PhD thesis. Jan. 2009 (cit. on p. 22).

[89]   B. Rafaely, B. Weiss, and E. Bachmat. "Spatial aliasing in spherical microphone arrays." In: *Signal Process. IEEE Trans.* 3 (2007) (cit. on p. 24).

[90]   I. Sloan and R. Wommersly. "Constructive Polynomial Approximation on the Sphere." In: *J. Approx. Theory* 103 (2000), pp. 91–118. ISSN: 00219045. DOI: 10.1006/jath.1999.3426 (cit. on p. 24).

[91]   M. J. Mohlenkamp. "A fast transform for spherical harmonics." In: *J. Fourier Anal. Appl.* 5.2-3 (1999), pp. 159–184. ISSN: 1069-5869. DOI: 10.1007/BF01261607 (cit. on p. 24).

[92]   J. Klein. "Optimization of a Method for the Synthesis of Transfer Functions of Variable Sound Source Directivities for Acoustical Measurements." Diploma Thesis. 2012 (cit. on p. 24).

[93]   M. Pollow et al. "Calculation of Head-Related Transfer Functions for Arbitrary Field Points Using Spherical Harmonics Decomposition." In: *Acta Acust. united with Acust.* 98.1 (Jan. 2012), pp. 72–82. ISSN: 16101928. DOI: 10.3813/AAA.918493 (cit. on p. 25).

[94]   R. Duraiswami, D. N. Zotkin, and N. A. Gumerov. "Interpolation and range extrapolation of HRTFs." In: *IEEE Int. Conf. Acoust. Speech, Signal Process.* (2004) (cit. on p. 25).

[95]   I. Ben Hagai et al. "Acoustic centering of sources measured by surrounding spherical microphone arrays." In: *J. Acoust. Soc. Am.* 130.4 (2011), pp. 2003–2015. ISSN: 0001-4966. DOI: 10.1121/1.3624825 (cit. on p. 26).

[96]   B. S. Masiero. *Individualized Binaural Technology*. ISBN: 9783832532741 (cit. on pp. 28, 58, 113).

[97]   J. Wenderoth. "Aktive Korrektur der Richtcharakteristik eines Lautsprecherarrays." Bachelorthesis. RWTH Aachen University, 2014 (cit. on p. 28).

[98] S. Müller. "Digitale Signalverarbeitung für Lautsprecher." PhD thesis. 1999 (cit. on p. 31).

[99] M. Pollow et al. "Fast measurement system for spatially continuous individual HRTFs." In: *4th Int. Symp. Ambisonics Spherical Acoust.* University of York, UK, 2012 (cit. on p. 33).

[100] M. Guski. "Martin Guski Influences of external error sources on measurements of room acoustic parameters." Dissertation. RWTH Aachen University. ISBN: 9783832541460 (cit. on p. 39).

[101] J.-G. Richter, G. Behler, and J. Fels. "Evaluation of a fast HRTF measurement system." In: *AES 140th Conv.* (2016), p. 9498 (cit. on p. 40).

[102] J. C. Middlebrooks. "Individual differences in external-ear transfer functions reduced by scaling in frequency." In: *J. Acoust. Soc. Am.* 106.3 Pt 1 (1999), pp. 1480–1492. ISSN: 00014966. DOI: 10.1121/1.427176 (cit. on p. 40).

[103] F. Zotter. "Sampling strategies for acoustic holography/holophony on the sphere." In: *NAG-DAGA, Rotterdam* (2009) (cit. on pp. 41, 107).

[104] J. Keiner, S. Kunis, and D. Potts. "Efficient reconstruction of functions on the sphere from scattered data." In: *J. Fourier Anal. Appl.* 13.4 (2007), pp. 435–458. ISSN: 10695869. DOI: 10.1007/s00041-006-6915-y (cit. on p. 41).

[105] F. P. Mechel. *Formulas of acoustics.* Springer Science & Business Media, 2013 (cit. on p. 45).

[106] K. Iida and Y. Ishii. "3D sound image control by individualized parametric head-related transfer functions." In: *Inter-Noise 2011* (2011), pp. 1–8 (cit. on p. 51).

[107] M. Morimoto and H. Aokata. *Localization cues of sound sources in the upper hemisphere.* 1984. DOI: 10.1250/ast.5.165 (cit. on p. 60).

[108] T. Van den Bogaert et al. "Horizontal localization with bilateral hearing aids: Without is better than with." In: *J. Acoust. Soc. Am.* 119.1 (2006), p. 515. ISSN: 00014966. DOI: 10.1121/1.2139653 (cit. on p. 62).

[109] W. R. Thurlow and J. R. Mergener. "Effect of stimulus duration on localization of direction noise stimuli." In: *J. Speech Hear. Res.* 13.4 (1970), pp. 826–838. ISSN: 00224685 (cit. on pp. 63, 70).

[110] J. A. Pedersen and P. Minnaar. "Evaluation of a 3D-audio system with head tracking." In: *AES 120th Conv.* (2006) (cit. on p. 63).

[111] J. Oberem et al. "Experiments on localization accuracy with non-individual and individual HRTFs comparing static and dynamic reproduction methods." In: *Proc. 44th DAGA*. 2018, pp. 702–705 (cit. on pp. 67, 69).

[112] B. F. G. Katz and M. Noisternig. "A comparative study of Interaural Time Delay estimation methods." In: *J. Acoust. Soc. Am.* 135.6 (June 2014), pp. 3530-40. ISSN: 1520-8524. DOI: 10.1121/1.4875714 (cit. on p. 70).

[113] R. B. King and S. R. Oldfield. "The impact of signal bandwidth on auditory localization: Implications for the design of three-dimensional audio displays." In: *Human factors* 39.2 (1997), pp. 287–295 (cit. on p. 70).

[114] D. Begault. *3-D Sound for Virtual Reality and Multimedia.* Vol. 19. 4. 2000, p. 99. ISBN: 0120847353. DOI: 10.2307/3680997 (cit. on p. 75).

[115] J. Richter and J. Fels. "On the Influence of Continuous Subject Rotation During High-Resolution Head-Related Transfer Function Measurements." In: *IEEE/ACM Transactions on Audio, Speech, and Language Processing* 27.4 (Apr. 2019), pp. 730–741. ISSN: 2329-9290. DOI: 10.1109/TASLP.2019.2894329 (cit. on p. 77).

[116] B. Kollmeier, T. Brand, and B. Meyer. "Perception of speech and sound." In: *Springer Handb. speech Process.* Springer, 2008, pp. 61–82 (cit. on p. 79).

[117] E. G. Shower and R. Biddulph. "Differential Pitch Sensitivity of the Ear." In: *J. Acoust. Soc. Am.* 3.2A (1931), pp. 275–287. ISSN: 0001-4966. DOI: 10.1121/1.1915561 (cit. on p. 79).

[118] P. Dietrich. "Uncertainties in Acoustical Transfer Functions." PhD thesis. RWTH Aachen University, 2013, p. 203. ISBN: 9783832535513 (cit. on p. 80).

[119] D. Schröder. "Physically based real-time Auralization of interactive Virtual Environments." PhD thesis. RWTH Aachen University, 2011, p. 207. ISBN: 9783832530310 (cit. on p. 87).

[120] D. Osterhammel and P. Osterhammel. "High-frequency audiometry: Age and sex variations." In: *Scand. Audiol.* 8.2 (1979), pp. 73–80. ISSN: 01050397. DOI: 10.3109/01050397909076304 (cit. on p. 87).

[121] V. R. Algazi, C. Avendano, and D. Thompson. "Dependence of subject and measurement position in binaural signal acquisition." In: *J. Audio Eng. Soc.* 47.11 (Nov. 1999), pp. 937–947. ISSN: 10052755 (cit. on p. 93).

[122] J. He, R. Ranjan, and W. S. Gan. "Fast continuous HRTF acquisition with unconstrained movements of human subjects." In: *ICASSP, IEEE Int. Conf. Acoust. Speech Signal Process. - Proc.* 2016-May (2016), pp. 321–325. ISSN: 15206149. DOI: 10.1109/ICASSP.2016.7471689 (cit. on p. 93).

[123] K. A. J. Riederer. "Part Va : Effect of Head Movements on Measured Head-Related Transfer Functions." In: (2004), pp. 795–798 (cit. on p. 93).

[124] T. Hirahara et al. "Head movement during head-related transfer function measurements." In: *Acoust. Sci. Technol.* 31.2 (Mar. 2010), pp. 165–171. ISSN: 1347-5177. DOI: 10.1250/ast.31.165 (cit. on p. 94).

[125] T. Carpentier et al. "Measurement of a head-related transfer function database with high spatial resolution." In: *Forum Acoust.* c (Sept. 2014), pp. 1–6. ISSN: 22213767 (cit. on p. 94).

[126] F. Denk et al. "Controlling the Head Position during individual HRTF Measurements and its Effect on Accuracy Headtracking and Visual Feedback for Head Positioning Control Head Positioning Results." In: (2017), pp. 1085–1088 (cit. on pp. 94, 95, 98, 99).

[127] S. Wepner, J.-G. Richter, and J. Fels. "Subject movement during the measurement of head-related transfer functions." In: *Fortschritte der Akust. DAGA 2019.* 2019 (cit. on p. 94).

[128] S. Zillekens. "Messung von individuellen HRTFs und Nachbereitung mittels Kugeloberflächenfunktionen Measurement of individual HRTFs and post-processing using spherical harmonics decomposition." Diploma Thesis. 2014 (cit. on pp. 94, 104).

[129] S. Wepner. "Influence of subject movement on the measurement of individual Head-Related Transfer Functions." Masterthesis. RWTH Aachen University, 2018 (cit. on pp. 96, 97, 99).

[130] R. Nicol et al. "A Roadmap for Assessing the Quality of Experience of 3D Audio Binaural Rendering." In: *Proc. EAA Jt. Symp. Auralization Ambisonics.* 2014, pp. 100–106. ISBN: 0007-0920. DOI: 10.1038/bjc.2011.405 (cit. on p. 116).

[131] R. Bomhardt, M. de la Fuente Klein, and J. Fels. "A high-resolution head-related transfer function and three-dimensional ear model database." In: 050002.2016 (2016), p. 050002. ISSN: 1939800X. DOI: 10.1121/2.0000467 (cit. on p. 116).

[132] R. Sridhar, J. G. Tylka, and E. Y. Choueiri. "A database of head-related transfer function and morphological measurements." In: *Proc. 143rd Conv. Audio Eng. Soc.* Audio Engineering Society. 2017, pp. 1–5 (cit. on p. 116).

Curriculum Vitae

Personal Data

	Jan-Gerrit Richter
15.07.1986	born in Bochum, Deutschland

Education

| 1996–2005 | Hellweg Gymnasium, Bochum |
| 1992–1996 | Dietrich-Bonhoeffer-Schule, Bochum |

Higher Education

| 2006–2012 | Dipl.-Ing. (Master Degree) Computer Engineering (Electrical Engineering and Information Technology) RWTH Aachen University |

Work Experience

| 2013 – 2018 | Research Assistant Institute of Technical Acoustics, Medical Acoustics Group RWTH Aachen University |
| 2011 | Six month internship, National Research Council (NRC) Institute for Research in Construction Ottawa, Canada |

April 2, 2019

Bisher erschienene Bände der Reihe

Aachener Beiträge zur Akustik

ISSN 1866-3052
ISSN 2512-6008 (seit Band 28)

Alle erschienenen Bücher können unter der angegebenen ISBN-Nummer direkt online
(http://www.logos-verlag.de) oder per Fax (030 - 42 85 10 92) beim Logos Verlag
Berlin bestellt werden.